The Bigger Bang

Societies through the ages have always been fascinated with our origins. In the last few years, scientists have begun to answer some of the most fundamental questions about the origin and early evolution of the universe. This book presents a fresh, engaging and highly readable introduction to these ideas.

Using novel, down-to-earth analogies, author James Lidsey steers us deftly on a journey to the cutting edge of cosmology. Step-by-step, we travel back in time through Lidsey's book until we arrive at the very origin of the universe. There we look at the fascinating ideas scientists are currently developing to explain what happened in the first billion, billion, billion, billionth of a second of the universe's existence – the 'inflationary' epoch. Along the way, we are given lucid accounts of many fascinating topics in theoretical cosmology, including the latest ideas on superstrings, parallel universes, and the ultimate fate of our universe. We also discover how the world of the very small (described by the physics of elementary particles) and the world of the very large (described by cosmology) are inextricably linked by events which wove them together in the first few moments of the universe's history.

Lucid analogies, clear and concise prose and straightforward language make this book a delight to read. It makes accessible to the general reader some of the most profound and complex ideas about the origin of our universe currently vexing the minds of the world's best scientists.

James E. Lidsey is a Royal Society University Research Fellow at Queen Mary and Westfield College, University of London. His research interests focus on the very early universe, especially inflation and the cosmological aspects of superstring theory. In 1998, he appeared in the *Sunday Times* "Hot 100" list of promising academics. For recreation, he is learning to play the mandolin, but with limited success to date.

The Bigger Bang

James E. Lidsey
Queen Mary and Westfield College
University of London

CAMBRIDGE
UNIVERSITY PRESS

PUBLISHED BY THE PRESS SYNDICATE OF THE UNIVERSITY OF CAMBRIDGE
The Pitt Building, Trumpington Street, Cambridge, United Kingdom

CAMBRIDGE UNIVERSITY PRESS
The Edinburgh Building, Cambridge CB2 2RU, UK
40 West 20th Street, New York, NY 10011-4211, USA
10 Stamford Road, Oakleigh, VIC 3166, Australia
Ruiz de Alarcón 13, 28014 Madrid, Spain
Dock House, The Water Front, Cape Town 8001, South Africa

http://www.cambridge.org

First published 2000

Printed in the United States of America

Typefaces Sabon 10/13 pt. and Antique Olive *System* LaTeX 2_ε [TB]

A catalog record for this book is available from the British Library.

Library of Congress Cataloging in Publication Data

Lidsey, James E., 1967–

The bigger bang / James E. Lidsey.

p. cm.

ISBN 0-521-58289-X (hb)

1. Cosmology. I. Title.

QB981 .L536 2000
523.1′8 – dc21

00-027639

ISBN 0 521 58289 X hardback

Contents

Preface

We live in a big universe. Even if we were able to travel across the universe at the speed of light, the journey would take us at least ten billion years. Why is the universe so large? Has the universe always been this big, or was it smaller in the past? If smaller, how small was it? Was there a time when the volume of the universe vanished?

We can ask related questions regarding matter in the universe. Why is the universe not empty? From where do the atoms that make up our bodies originate? When were these atoms created?

Questions such as these lead us inevitably to the origin of the universe. Did the universe have a definite beginning, or has it always existed? If it had a beginning, can we talk meaningfully about what might have happened beforehand? And what caused the universe to come into existence in the first place?

The purpose of this book is to address questions such as these. Moreover, because our own origin is linked with that of the universe as a whole, we are indirectly studying our own past when we investigate the beginning of the universe.

We will see that the structure of the universe is intimately related to the structure of the smallest elementary particles. This relationship between the world of the very large and that of the very small was manifest even during the first second of the universe's history. Remarkably, the conditions that prevailed when the universe was no more than a fraction of a second old may have led to the formation of galaxies, stars and planets. Our existence billions of years later depends directly on what happened at that very early time.

Throughout this book we will encounter very large and very small numbers. The standard notation is to express such numbers as powers of ten. Thus one million (1,000,000) is ten to the power six because there are six zeros that follow the 1. It is written as 10^6. One billion (one thousand million), then, is written as 10^9. We will refer to one million million as one trillion and write it as 10^{12}. Very small numbers are written in a similar way. For example, one millionth is one divided by a million and is written as 10^{-6}. One billionth is denoted by 10^{-9}, and so on.

We will also encounter in this book references to a wide range of temperatures. Unless otherwise stated, we will measure temperature in degrees Celsius. The lowest temperature possible is −273.16°C, which is known as *absolute zero*. The temperature of outer space, for example, is about three degrees above absolute zero.

Acknowledgements

It is a pleasure to thank: Tony Mason and Andrew Liddle for their encouragement during this book's early stages; Terry Arter, Richard Frewin and Eamon Kerrins for help in preparing the figures; Peter Coles and Amitabha Lahiri for useful discussions; and Jarvis Brand, Nick Hill, John Lidsey, Keith Myles and Reza Tavakol for reading through the manuscript. I would also like to thank Barbara for her unlimited friendship.

1

The Structure of the Universe

During the past few decades of research a plausible picture of the universe's most distant past has begun to emerge. The current view is that the universe came into existence some ten billion years ago in the form of a huge, exploding 'fireball'. This was the *big bang.*

We are going to discuss some key features of the big bang in this book. In particular, we will look at the central question of just how 'big' it really was. But before we can begin our journey back towards the origin of the universe, we must work out our present location within it. Let us therefore embark on a brief sight-seeing tour of the universe.

The Earth belongs to a collection of objects known as the *solar system.* The central and largest object in this system is the sun. Nine planets, including the Earth, orbit the sun. Pluto is the planet most distant from the sun, and Pluto's orbit may be viewed as the edge of the solar system.

The nearest significant object to the Earth, its moon, is some four hundred thousand kilometres away. For comparison, the distance between the Earth and the sun is roughly one hundred and fifty million kilometres, whereas the average distance between the sun and Pluto is approximately six billion kilometres.

What lies beyond the solar system? As we travel past Pluto, the vastness of empty space soon becomes apparent. For example, the nearest star to the sun – Proxima Centauri – is some forty trillion kilometres away from it.

We thus encounter a major problem even during this early stage of our journey: When thinking about the universe, how are we to deal with the huge distances involved? We were already talking of billions of kilometres before we left the confines of the solar system. This distance is in itself difficult to imagine. Yet we must increase this scale to trillions of kilometres before we arrive at the nearest star.

Astronomers make sense of such large scales by measuring distance in terms of a *light year*. This is the distance light travels in one year when moving at its speed of three hundred thousand kilometres per second. Numerically, one light year is equivalent to nine and one-half trillion kilometres. The distance from the Earth to the sun is about eight light minutes. This is the time it takes light to travel from the surface of the sun to us here on Earth. The distance from the sun to Proxima Centauri is over four light years.

Scale models also prove useful when comparing distances between various objects in the universe. Let us consider what would happen if we were to reduce all distance scales so that the radius of the Earth was comparable to the radius of a typical wristwatch. In this case, the radius of the sun would be equivalent to the height of an average man. The distance between the Earth and the sun would then be four hundred metres. Pluto would be some fifteen kilometres away, but we would still have to travel about one hundred thousand kilometres before we reached Proxima Centauri. Such a trip would be equivalent to travelling two and one-half times around the world.

Our sun and Proxima Centauri are just two of the many stars that belong to the Milky Way Galaxy. If we were to move outside our galaxy and view it from above, we would find that it looks rather like a giant Catherine-wheel, as shown in Figure 1.1a. The Milky Way contains a number of spiral arms that are attached to a central region. These arms consist of numerous stars. When viewed from its side, the galaxy resembles a disc with a bulge in its centre, as shown in Figure 1.1b. The radius of the bulge is about ten thousand light years, and the disc itself is at least one hundred thousand light years across. The disc can be seen on a clear, moonless night and resembles a thin cloud that stretches across the sky. It has a diffuse appearance, and, indeed, the word 'galaxy' can be traced to the Greek word *galacticos*, which means milky.

There is also a halo of very old stars around the centre of the galaxy. This halo extends out in all directions for about fifty thousand light

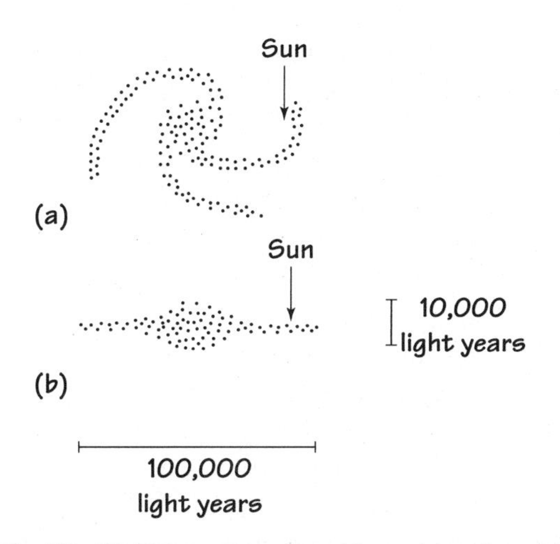

Figure 1.1. *(a)* The Milky Way Galaxy when viewed from above. It contains a spherical, central region and a number of spiral arms. These consist of stars. The location of the sun near one of these arms is shown. The sun lies about 28,000 light years from the centre of the galaxy. *(b)* The galaxy when viewed from its side. At its widest point, the galaxy's width is about 100,000 light years.

years. In total, the Milky Way contains over one hundred billion stars. Our sun is located about 28,000 light years from the centre of the galaxy.

We need a further reduction of scale to make comparisons on the galactic scale. Let us shrink the entire solar system so that it has a size comparable to that of a typical grain of sand. (Recall that the solar system's actual size is about six billion kilometres.) The nearest star, Proxima Centauri, would now be just over one metre away from the edge of the solar system. The distance from the solar system to the centre of the galaxy would correspond to the height of Mount Everest. When comparing the solar system to the rest of the galaxy, we can think of a mountaineer who has reached the summit of Everest and in whose pocket is a grain of sand.

The overall picture we have thus far is that the sun and other stars in our galaxy are separated by many thousands of light years. Yet despite these great distances, the stars are still attracted to one another by the force of gravity. It is this attraction that keeps the stars confined to the galaxy.

What do we find when we move beyond the neighbourhood of the Milky Way? As far as our journey through the universe is concerned, we have only just begun. Once more we are confronted with the vastness of empty space. We do not encounter another significant object until we have travelled outwards for a further 170,000 light years. At this distance, we find a small galaxy known as the Large Magellanic Cloud.

There are numerous other galaxies in the universe besides the Large Magellanic Cloud and the Milky Way. The observable universe probably contains over one hundred billion galaxies. These galaxies come in many shapes and sizes. Many are spiral in shape just like the Milky Way, although the majority are not. Those that are not are referred to as 'elliptical' galaxies due to their shape. These elliptical galaxies are dominated by stars that may be as old as ten billion years.

Although galaxies exist as separate entities throughout the universe, they do not behave as isolated objects. They attract each other by the force of gravity and so group together into clusters. The number of galaxies in a particular cluster may be quite low, but can be as high as a few thousand. Typically, a cluster of galaxies extends for millions of light years. For example, our own Milky Way belongs to a cluster known as the *local group*. The largest galaxy in this group is the Andromeda Galaxy, a spiral galaxy that is over two million light years away from the Milky Way. The local group extends for about six million light years, about sixty times the size of our own galaxy.

Clusters of galaxies are grouped into superclusters. Superclusters extend for hundreds of millions of light years. Our local group of galaxies belongs to what is known as the *local supercluster*. At the centre of this supercluster is the cluster of galaxies known as the Virgo cluster. The Virgo cluster, which contains thousands of galaxies, is located about fifty million light years away from our own local group. In a broad sense, the universe may be viewed as a hierarchical structure of galaxies, clusters of galaxies and superclusters of galaxies.

Finally, the most distant visible objects that have been observed to date are known as *quasars*. Quasars emit an enormous amount of energy, but the source of this energy has not yet been identified. These objects are thought to be at least ten billion light years away from us, and this distance represents the size of the observable universe.

Some typical distance scales are summarized in Table 1.1. What have we discovered from this tour around the universe? We see that our planet orbits an average star that is located in the outer regions of the Milky Way Galaxy. Our galaxy contains at least one hundred billion stars and is just one example of the hundred billion galaxies that constitute the observable universe. Although we tend to think of the other planets in our solar system as being a great distance from us,

Table 1.1. *Some typical distance scales in the universe. The symbol 'ly' stands for light year and corresponds to 9.5 trillion kilometres*

Objects	Distance
Earth–Moon	4×10^5 km
Earth–Sun	1.5×10^8 km
Solar system diameter	6×10^9 km
Sun–Nearest star	$4.3ly$ (4×10^{13} km)
Milky Way diameter	$10^5 ly$ (10^{18} km)
Cluster of galaxies	$\geq 10^6 ly$ (10^{19} km)
Size of the universe	$\geq 10^{10} ly$ (10^{23} km)

the entire solar system is *tiny* when compared to our own galaxy, let alone to the rest of the universe.

Given this broad picture, we can proceed to investigate the structure and history of the universe in more detail. The study of the universe as a whole is known as *cosmology*. The primary aim of the cosmologist is to understand how the universe developed over time into its present state and then to predict how it might behave in the future. One question that has occupied many cosmologists in recent years is how the current structure of the universe was influenced by physical processes that operated during the big bang.

Cosmologists are able to look progressively further back in time by probing the depths of the universe. Light that originates from a very distant galaxy has to travel farther to reach us here on Earth than does light emitted from a relatively nearby galaxy. This means that it takes longer for light from a distant galaxy to complete its journey. Many of the galaxies that we observe are so far away that it has taken their light billions of years to reach our solar system. Thus the photographs that we take of these galaxies are not pictures of what they look like today, but rather are images of what they looked like in the past.

The light emitted from distant galaxies has certain characteristic features, as we shall see in Chapter 3. These features indicate that the galaxies are moving away from each other. This finding is very significant, because it implies that the universe as a whole is *expanding*. Indeed, our observations indicate that the universe has been expanding for at least ten billion years.

Let us study the implications of this expansion further. We can view physical distances in the universe in terms of a given distance between two fictitious particles. Suppose we were to consider a particle in our own Milky Way Galaxy and a second particle in the neighbouring Andromeda Galaxy. An expansion of the universe can then be interpreted as an increase in the distance separating these two particles. This is what we shall have in mind when we talk about an increase in the volume of the universe.

As we go back in time, the distance between our two particles must have been somewhat smaller than it is today. This provides us with our first insight into what the universe may have looked like at earlier times. It is reasonable to suppose that the universe must have been *smaller* at some stage in its history than it is at present. Consequently, the galaxies must have been closer together than they are today, and the density and temperature of matter must have been correspondingly higher. If we are prepared to go sufficiently far into the past, the distance separating our two particles would have been much smaller than the size of a typical galaxy. At these very early times, galaxies as we know them today could not have existed. All the matter in the universe would have behaved as though it were a superhot and superdense fluid.

The big bang model describes the universe when it went through this early phase, after it was just a few seconds old, and we summarize the key features of this model in Chapter 7. We will also consider what may have happened in the universe before the first second had elapsed. In Chapters 8 and 9 we will see that there are strong arguments to suggest that the universe underwent a period of very rapid expansion when it was no more than 10^{-35} seconds old. At that time the matter currently contained within the observable universe (around one hundred billion galaxies) would have been squashed into a region of space considerably smaller than that occupied by a typical atom. Furthermore, the temperature of the universe would have been exceptionally high, many times higher than the temperature at the centre of the sun.

This period of expansion is referred to as *inflation*, because the universe increased in size by a huge factor. It did this very quickly indeed. The duration of this rapid expansion was extremely brief, at least in the simplest versions of the theory, and may have taken less than 10^{-33} seconds to complete.

If inflation is to provide us with a plausible picture of the universe at these very early times, we need to understand what caused the universe to expand so rapidly. In the next five chapters we will develop the background necessary for discussing the very first moments of the universe's history. We will then proceed in the remainder of the book to discuss what we think may have happened during the first 10^{-35} seconds of the universe's existence. This will lead us to some of the more speculative ideas that have been developed recently regarding the origin of the universe.

Where is a suitable place to begin? It seems reasonable that we start from somewhere relatively close to home – that is, from somewhere within the solar system. Since the sun is the largest object in this system, we could begin by discussing its properties. The most obvious feature of the sun is that it shines. Let us begin then by asking what it is that causes the sun to shine.

2

Why Does the Sun Shine?

Visible light is an example of *electromagnetic radiation*. This radiation may be pictured as a wave travelling through space. Although light always travels at a fixed speed, its wavelength – defined as the distance between two successive peaks or troughs – is not uniquely specified. Different types of light can have different wavelengths. These differences manifest themselves as different colours in the visible spectrum. For example, red light has a slightly longer wavelength than blue light. The light that we receive from the sun is a mixture of all the different colours.

Electromagnetic radiation with wavelengths significantly longer or shorter than those associated with visible light also exists. Two examples are gamma rays and radio waves. All types of electromagnetic radiation carry a certain amount of energy. A gamma ray has a lot of energy whereas a radio wave carries a relatively small amount of energy. In a sense, we can imagine the energy as localized around the peaks and troughs of the wave. Thus the energy of a given type of electromagnetic radiation is specified by its wavelength; a shorter wavelength corresponds to a higher energy and vice versa. This follows since a shorter wavelength means that more crests and troughs will arrive in a given interval of time, so more energy will be received.

Light has a very important property in that it changes its direction of motion as it travels between regions of different density. Much of what we know today about the internal composition of stars is deduced directly from this property. In effect, as it passes from one medium to another, light becomes deflected from its original path. What makes this phenomenon so interesting to us is that the precise amount of deflection is determined by the wavelength of the light. For example, the deviation of red light differs from that of the shorter-wavelength blue light. As a result, red light and blue light follow *different* paths. A beam of light that is made up of radiation of different wavelengths will separate into its individual colours because of the change in density.

Sunlight is just such a mixture of different colours and can be separated in this fashion. In fact, this is precisely what happens when a rainbow appears in the sky during a shower. We see the rainbow as the sunlight passes from the atmosphere, through the denser water droplets, and back out again into the atmosphere.

An identical effect arises when sunlight is passed through a prism onto a piece of photographic film. Once the film has been developed, the resulting photograph resembles a picture of a rainbow. However, a closer look at the picture reveals dark bands. These regions are so thin that they look as if someone has drawn a vertical line on the photograph with a black pen. Lines that are much brighter in intensity can also be seen. What is causing these dark and bright lines to appear in the picture?

Let us concentrate on the dark lines. Because each colour in the picture corresponds to light of a certain wavelength, the existence of dark lines at specific colours implies that the light with that particular wavelength is missing from the sunlight directed through the prism. It is important to ask what might have happened to this radiation. That we do not see it in the photograph tells us that it failed to reach the Earth's surface for some reason. One possibility is that this particular light was absorbed by something during its journey from the sun to the Earth. But this is unlikely, because the region between our planet and the sun is basically empty. It is also unlikely that this light could have been absorbed by the Earth's atmosphere. A more plausible possibility is that the light was never emitted in the first place. The light could have been absorbed by the material in the sun before it had time to escape from the sun's surface.

Can we determine the nature of the stellar material that is responsible for this absorption? The answer is yes, we can, but before doing so we need to study the internal structure of atoms.

The atom is the fundamental building block of all matter including that contained within our own bodies. The typical size of an atom is about 10^{-10} metres. Every atom has a nucleus that is made up of tiny particles called 'protons' and 'neutrons'. These particles are packed together very tightly inside the nucleus. For the purposes of this discussion, we may think of them as minute billiard balls. They have a diameter that is roughly 10^{-15} metres. Protons and neutrons are similar to each other, but they are not identical. They have roughly the same mass, although the neutron is slightly heavier than the proton. The proton also carries a positive electric charge, whereas the neutron is electrically neutral. This means that the nucleus as a whole is positively charged.

Surrounding this nucleus are particles known as electrons. The electrons are smaller and lighter than the protons and neutrons, so most of the mass of an atom is concentrated inside the nucleus. Electrons have negative electric charge, and an atom has just enough electrons to ensure that the positive charge of the nucleus is precisely cancelled. The atom is therefore electrically neutral.

The electrons around the nucleus of an atom are not free to assume just any orbit. Electrons have a tendency to remain as near to the nucleus as possible (without actually falling into it), and some are able to get relatively close. If there are many electrons inside the atom, the region closest to the nucleus becomes occupied and inaccessible to the remaining electrons, and these particles are then forced to occupy larger and larger orbits. This means that the orbits of the electrons are *restricted*.

This is important because the energy of each electron in the atom is determined by its distance from the nucleus. Those electrons that are farther away have a higher energy. To see why this is, let us consider the simplest atom in existence. This is the hydrogen atom. In the hydrogen atom the nucleus contains a single proton and has one electron orbiting around it. If we want to separate these two particles, we have to overcome the attractive force that operates between them due to their opposite electric charge.

A certain amount of energy must be expended in overcoming this resistance to separation. Since energy must always be conserved in any physical process, the energy it costs us to separate the proton and electron must go somewhere. It cannot simply disappear. It becomes stored as 'potential' energy in the electron. The electron *gains* energy as it becomes separated from the proton and assumes a larger orbit. It follows that if the orbit of an electron around the nucleus is restricted, its energy must also be constrained.

The act of increasing the distance between a given electron and the nucleus is rather like that of carrying a heavy object such as a brick up a ladder. In the former case, we must transfer energy to the electron to overcome the electrostatic attraction between the electron and the nucleus. In the latter example, we must do work to overcome the attractive force of gravity. The brick gains potential energy as its height above the ground increases, just as the electron gains energy when it is moved away from the nucleus. Because of this close similarity, we can draw an analogy between the two pictures and view the brick as an electron and vice versa. The height of the brick above ground level then corresponds to the separation between the nucleus and the electron in the atom.

The height and energy of the brick increase by a certain well-defined amount as we climb the ladder. This increase is determined by the separation between the rungs of the ladder. If we need to rest during the climb, we cannot hover between the rungs. The potential energy of the brick is restricted in a similar way to that of the electron. We may also think of an electron inside the atom as being constrained to lie on the rungs of a fictional ladder. This 'atomic ladder' is not real in the sense that it physically exists, but envisioning it does help us to picture what is going on. As shown in Figure 2.1, the electrons must lie on the

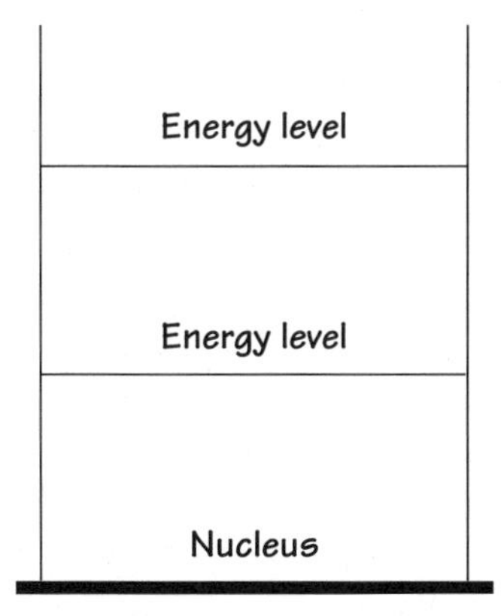

Figure 2.1. The allowed energy states of electrons inside atoms may be thought of as rungs on a ladder. The rungs represent energy levels. Electrons must sit on these rungs and cannot lie between them. In this picture, the nucleus of the atom sits somewhere below the bottom rung.

rungs and are forbidden from the regions between them. The nucleus can be found below the bottom rung.

Collectively, these rungs are referred to as 'energy levels', since they represent the amount of energy associated with the electrons. The minimum quantity of energy that an electron can have is determined by the closest possible orbit to the nucleus. This orbit corresponds to the lowest available rung on the ladder. Orbits that are farther away from the nucleus represent a higher energy and correspond to higher rungs on the ladder.

An electron would like to be on the bottom rung, but if this rung is already occupied, the electron must settle on a higher one. The problem the electrons have is similar to the one encountered by two people on a ladder when they attempt to balance on the same rung at the same time. There is not enough room for both of them!

This picture of electrons sitting on the rungs of a ladder holds for all chemical elements. It might be expected, though, that the allowed energies would be different for different elements, since each type of atom contains a different number of electrons. This is indeed the case. The *relative separation* between the rungs on the ladder is different for different elements. In other words, each type of atom – such as hydrogen, carbon and oxygen – has its own individual ladder associated with it.

This feature is important because it suggests that the structure of the ladders could be employed as a means of identifying the elements. We can think of the ladders as 'atomic signatures'. If we were able to measure the relative separation between the allowed energy levels of a given substance, it might be possible to identify which element was present.

How might we read the signature of an atom? Consider what happens when a hydrogen atom absorbs energy from an outside source. This is shown schematically in Figure 2.2. The energy could be in the form of electromagnetic radiation. Initially, the electron is located on the bottom rung of the ladder. When the energy is absorbed, it is transferred to the electron.

The energy of the electron is restricted, which means that the electron can absorb only a certain amount of energy. Increasing its energy causes the electron to jump to a higher rung on the hydrogen ladder. However, the electron desperately wants to get back to the bottom rung, because this is the state of lowest energy. In the same way, a brick will inevitably fall to the ground if released. After a very short time, therefore, the electron gives up its newly acquired energy and falls back down the ladder.

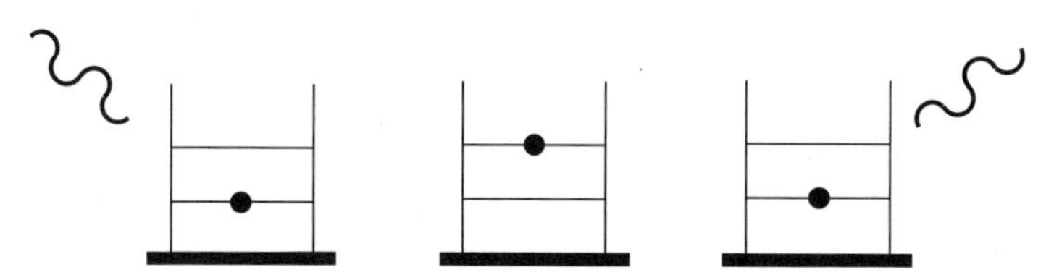

Figure 2.2. A packet of electromagnetic radiation is absorbed by the electron in a hydrogen atom. This causes the electron to jump to a higher rung on the ladder. After a short time, the electron loses this extra energy and falls back to the lower rung. The energy is released in the form of electromagnetic radiation.

As the electron falls, energy is released from the atom as electromagnetic radiation. The electron may fall straight to the level it was on originally in which case the energy of the emitted radiation is equal to the amount that was initially absorbed. Alternatively, the electron may fall to intermediate levels. The emitted radiation will then have different wavelengths to that of the absorbed light.

Since the amount of absorbed energy is fixed by the separation between the rungs, the energy of the emitted radiation will also be fixed. It follows that the possible wavelengths of the emitted radiation will be restricted.

In many instances the wavelength of the absorbed and emitted radiation can be in the visible spectrum. That is, some atoms can absorb and emit light of a specific colour. This feature leads us to consider the following scenario, as is shown in Figure 2.3. Suppose we take a pure element such as hydrogen and fire at it a beam of light that consists of many different wavelengths and colours. The hydrogen atoms will

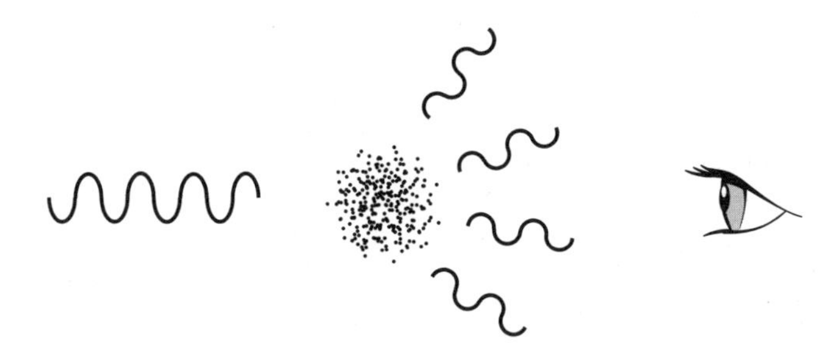

Figure 2.3. A beam of electromagnetic radiation of many different wavelengths approaches a cloud of hydrogen gas from the left. Most of the radiation is not absorbed and passes through to the observer on the right. Radiation with just the correct wavelength to cause electrons to jump to higher energy levels is absorbed. Although the radiation is quickly reemitted, it travels out of the cloud in a random direction and does not reach the observer.

absorb only the light that has the appropriate energy and wavelength. Radiation with other wavelengths will pass through the substance unaffected.

As the excited hydrogen electrons fall back down to the lowest available rungs the absorbed energy is quickly reemitted. In general, the emitted radiation has a different wavelength from the radiation that was absorbed. It also travels off in a random direction and does not follow the rest of the radiation. This implies that the intensity of certain wavelengths, or equivalently, colours, in the beam will be significantly reduced once the light has travelled through the hydrogen. If the beam is then sent through a prism and aimed at photographic film, all the colours in the resulting picture will be separated. The absence of certain wavelengths will appear as *dark* lines on the photograph. The crucial point is that the locations of these dark lines are determined by the wavelengths of the missing light, and these wavelengths in turn are determined by the amount of energy that can be absorbed. The wavelengths depend directly on where the rungs are situated on the ladder.

This is how the signature of an atom is read. The process can be repeated for all elements, and since each substance has its own unique set of rungs, different colours will be missing for different substances. The location of the dark lines in the photograph will be different in each case.

Atomic signatures determined in this fashion are very relevant to the study of stars. This technique enables us to identify the types of atoms present inside a star. Recall that the light from the sun contains dark lines in various places, thereby implying that sunlight of certain colours is missing. We suggested that these colours may have been absorbed by elements inside the star. If so, the absorbed light should reveal the identity of the elements responsible for the absorption. In principle, all we need to do is measure where the dark lines appear in the sunlight and then compare the results to the atomic signatures obtained in the laboratory. A match tells us that a given element is present.

This technique is not restricted to the sun and has been carried out on a large number of stars. By employing this method researchers have found that most stars, including the sun, are made up predominantly of hydrogen and helium. Hydrogen is the simplest element in existence. Its nucleus consists of a single proton. The helium nucleus is slightly more complicated and contains two protons and two neutrons.

Now that we have a handle on their composition, we may understand what causes stars such as the sun to shine. It is thought that the sun formed initially when a region of hydrogen gas began to collapse under its own gravity. As the collapse gathered pace, the hydrogen atoms in the centre of the cloud had less room in which to move around. They collided into each other more frequently and with greater and greater ferocity, causing the gas to become progressively hotter. The electrons in the hydrogen atoms were knocked about a great deal, and this made them jump to higher and higher energy levels. As the density of the central region continued to increase, the energies of the atoms became so high that the electrons were effectively held in these excited states.

The electrons in these hydrogen atoms were so energetic that they were unable to drop back down to the lowest energy state. If for some reason an electron did manage to fall to a lower level, it soon acquired new energy and was kicked back up the ladder. As the collapse proceeded, the electrons gained more and more energy and continued to climb up the ladders of energy levels. Eventually, they gained so much energy that they reached the top of the ladders. By this time the electrons had moved so far away from the nuclei that they were no longer held captive by them. They had effectively escaped and were now behaving as free particles. At this point, the hydrogen cloud began to resemble a dense, hot soup of negatively charged electrons and positively charged protons. This state of matter is known as a *plasma*.

The collapse of the hydrogen gas cloud continued, and the core heated up even further. The protons were free to collide directly with each other, and they did so with great enthusiasm. The density of the plasma eventually became so great that the protons had enough energy to overcome the repulsive effects of their own electric charges. Eventually, the temperature at the centre of the cloud reached ten million degrees, and something significant happened. The protons began colliding into one another with such force that they literally fused together to form helium nuclei. (More specifically, some of the protons decayed into neutrons and other particles. The newly created neutrons then combined with the protons to form helium.)

At this point the cloud has become a star. The actual process is quite complicated, but we need not go into the details here in order to develop a working picture of what happens. The essential point is that the helium nucleus, which consists of two neutrons and two protons, has a *lower* energy than do four unbounded protons. It is a fundamental principle that a physical system will always settle into the lowest energy state that is available to it, which is why a ball will freely roll down rather than up a hill. Consequently, it becomes energetically favourable for two protons in the core to convert into two neutrons and to combine with a further two protons to make a helium nucleus. This fusion of hydrogen into helium is occurring today in the centre of stars such as the sun.

Since the resulting helium nucleus has a lower energy than the original four protons, energy must have been released during the fusion process. In fact, an enormous amount of energy is released as electromagnetic radiation. This radiation has a tendency to move away from the core of the star towards the surface. In contrast, the protons, which are attracted towards the centre by the force of gravity, prevent most of the radiation from escaping.

We may picture the radiation and the protons as two separate crowds of people that are travelling in opposite directions through a very narrow alley. Once they meet, the two groups jostle each other and soon become stuck in the confusion. Neither group is then able to make further progress because of the influence of the other. The progress of one group is effectively cancelled out by the action of the other. In a similar way, in the star, the force of the radiation moving outwards and the inward force of the protons balance each other.

This interplay between the protons and the radiation prevents further collapse of the star. Consider what would happen if the collapse were to continue. The density of the matter would rise, and this would make the protons even more energetic than they already are. The protons would be able to fuse together more easily, producing more helium and radiation. The outward pressure exerted by the radiation would increase, and this would eventually come to dominate the inward pull of gravity. The protons would be forced outwards from the core, and the star would expand slightly. But as it did so, the star would cool, and the energy of the protons would fall. The fusion of helium, together with the subsequent release of radiation, would become less efficient. If the star expanded too far, an insufficient amount of radiation would be produced, and this would allow collapse to proceed once more.

This interplay between the matter and radiation results in an equilibrium being established; the outward pressure of the radiation *exactly* counterbalances the inward pull of gravity due to the matter. This state of affairs is extremely stable, and further collapse of the star is prevented as long as sufficient radiation is produced in the centre. The size of the star remains essentially constant during the fusing of hydrogen into helium. This equilibrium explains why the sun is so stable and is not either collapsing in on itself or exploding.

Nevertheless, a very small fraction of the radiation produced in the core is able to escape from the centre of the sun and reach its surface. Some of this electromagnetic radiation is in the form of visible light, and some of this light eventually finds its way to the surface of the Earth. It is this light that we see during the day.

We have therefore answered the question we posed at the end of the previous chapter. A star such as the sun spends most of its life converting hydrogen into helium. The amount of hydrogen that is available for fusing is limited, however, and the supply will eventually become exhausted. Although our sun should continue this process for another five billion years or so, eventually the conversion of hydrogen into helium will no longer occur, and the release of radiation will stop. In the absence of this radiation, there will be nothing to prevent the sun from collapsing further. The consequences of this secondary collapse depend on the size of the star, and we shall defer our discussion of what happens until later.

The technique of identifying atomic structure that we have discussed in this chapter may also be employed to study the light arriving from distant galaxies. This will be the topic of the next chapter.

3

The Expansion of the Universe

If the stars in distant galaxies have the same composition as those in our own galaxy, the light we receive from them should exhibit the characteristic signatures of hydrogen and helium. This is indeed observed. However, there is one crucial difference between galactic and stellar light, and this has profound consequences for our understanding of the universe. Although it is true that the *relative* separation of the dark lines observed in the galactic light corresponds to hydrogen and helium atoms, the lines do not appear quite in their expected positions. When we examine the light from galaxies, we find that the dark lines have all been shifted slightly along the photograph.

Figure 3.1 shows schematic pictures of the light emitted by a typical star in our galaxy and the light received from an average distant galaxy. Light of a longer wavelength is positioned towards the right of the diagram. If this were a colour photograph, the picture would appear redder on the right-hand side and bluer on the left. Notice how the absorption lines in the galactic light are all positioned at slightly *longer* wavelengths than in the stellar case.

The key to understanding what causes the dark lines to be moved towards longer wavelengths lies in the fact that light travels through space as a wave. Consider what happens when a wave is emitted by a source and is then picked up by a receiver positioned some distance away. If the emitter and receiver remain at rest relative to each other,

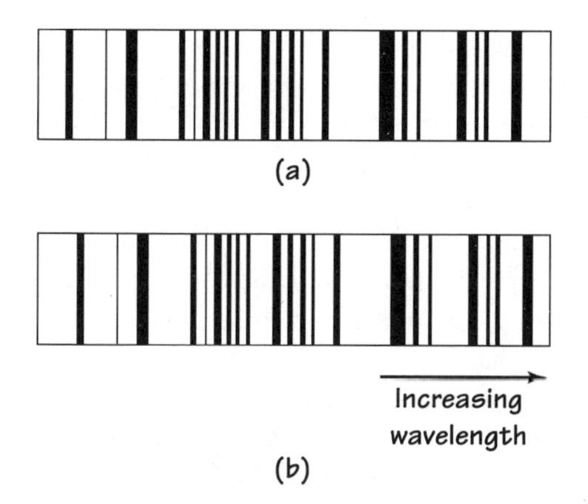

Figure 3.1. *(a)* A schematic picture of the light received from stars in our own galaxy. *(b)* The corresponding spectrum of light received from a distant galaxy. In both cases, the spectrum contains a series of dark lines due to the absorption of light of certain wavelengths. The relative separation of the lines is the same in both cases, but the lines in the spectrum of the distant galaxy are shifted to slightly longer wavelengths. This implies that the galaxy is moving away from the Milky Way.

the distance that separates them stays constant with time. Hence, each crest of the wave takes the same amount of time to reach the receiver, and this means that an equal number of crests will arrive during a given time interval. The wavelength of the wave will be the same at both the emitter and the detector.

This is *not* the case if the emitter is moving away from the detector. If the distance between the emitter and receiver increases with time, each crest has to travel a little farther than the one that went before it, and it will take slightly longer for each successive crest to arrive at the receiver. This will result in a small *increase* in the distance between two neighbouring crests. In other words, the wavelength as measured at the receiver will be slightly *longer* than that measured at the emitter. The overall effect of the change in the relative separation of the emitter and receiver is to increase the observed wavelength of the wave.

This shift in the wavelength is known as the *Doppler effect* after Christian Doppler, the Austrian physicist who discovered it. The familiar example of a whistle sounded from a moving train illustrates the point. As the train approaches the station, an observer on the platform detects an increase in the pitch of the whistle because the frequency of the note is rising. The pitch then falls once the train has passed through the station because the train is moving away from the platform.

Since light has wavelike properties, its wavelength also increases if the emitter and detector are moving away from each other. We can think of light as being emitted by the galaxies at some point in the distant past and later received here on Earth by our telescopes. What would happen if all the galaxies were moving away from us? We would expect the wavelengths of their light to be increased due to the Doppler effect. The absorption lines would then be located at slightly larger wavelengths than we would expect to see if the galaxies were at rest. This is precisely what is shown in Figure 3.1.

Discovered in the 1920s, this shifting of galactic light can be explained by the motion of galaxies *away* from ours. This is a fundamental cosmological observation, and it leads us to conclude that the universe is *expanding*.

What do we mean when we say that the universe is expanding? From our vantage point here on Earth, it appears that all the other galaxies in the universe are moving away from ours. When we attempt to imagine this state of affairs in our mind, we naturally think of our galaxy as being positioned at some fixed location in space. In this picture, the rest of the universe is moving away from us in much the same way that debris rushes away from the centre of an explosion.

We should stress immediately that because it places our galaxy and ourselves at the very centre of the universe this picture is *not* correct. When we do cosmology, we have to assume that there is nothing privileged or unique about our position in the universe. How, then, should we interpret this observed motion of the galaxies away from ours? We may consider the following possibility. Suppose that we were able to travel anywhere within the universe in a matter of moments. As we hopped from one galaxy to another, what would we see? If there is nothing special about our current position in the universe, what we would see should be roughly the same anywhere. It follows that the universe should appear similar to an observer using a telescope *regardless* of where the telescope happens to be placed. As we travelled through the universe, we would expect to see the same picture as we see here on Earth. That is, we would see all galaxies in the universe moving away from the particular one in which we happened to be located.

This is all very well, but how does it help us to picture the expansion of the universe in a realistic way? Let us suppose that we return home and persuade a friend to make a similar journey to a nearby galaxy. We both agree to measure the separation speed of the two galaxies in which we are situated and also to investigate some other galaxies

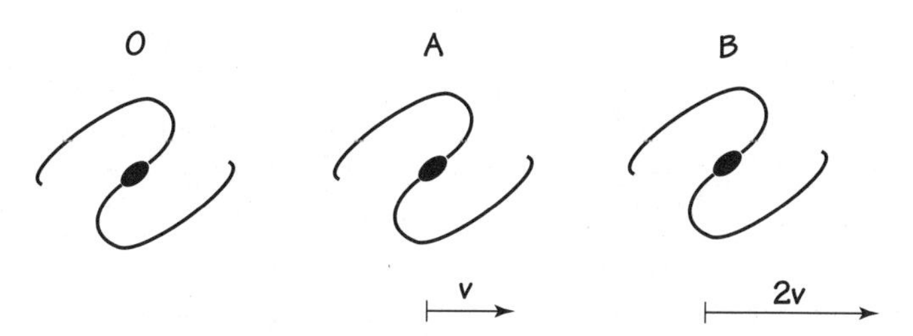

Figure 3.2. When measured from galaxy *O*, the apparent velocity *v* of a galaxy grows in direct proportion to its separation distance.

that are nearby. This arrangement is shown in Figure 3.2. We have labelled our galaxy as *O* and the galaxy to which our friend travels as *A*. Another galaxy, denoted by *B*, is also shown. We assume, for simplicity, that all the galaxies are evenly separated. Thus, galaxy *B* is twice as far from *O* as galaxy *A*.

We receive light from galaxy *A* and deduce that the galaxy is moving away from us with a particular speed, say one hundred kilometres per second. There is supposed to be nothing special about our position or about any observations that we might make. This means that our friend must observe galaxy *B* to be moving away from him at the same speed of one hundred kilometres per second, since the separation between *A* and *B* is the same as that between *O* and *A*. Galaxy *B* must be moving away from us at twice this speed because it is twice as far away from us as galaxy *A*.

This observation implies that the apparent separation speed of a galaxy from the observer increases in *direct proportion* to its distance. This relationship was discovered experimentally by the American astronomer Edwin Hubble in 1929. It is a direct consequence of the assumption that our position in the universe is equivalent to any other.

We conclude from this chapter that the universe is expanding in the sense that the average distance between galaxies is increasing over time. No single galaxy is stationary, and all galaxies move with respect to one another. One consequence of this observation is that in earlier times the universe must have been a lot smaller than it is today. Indeed, if this expansion has proceeded throughout the history of the universe, the universe must have been extremely small at very early times. It would have been so small, in fact, that the galaxies would have effectively been squashed together on top of one another.

How far back in time do we need to go to get to this point? If we know the distance to the furthest galaxies and also know the speed at which they are receding from us, the time that has elapsed since they were very close is simply given by the distance divided by the separation speed. In a sense, this elapsed time represents the age of the universe. The observable universe is roughly ten billion light years in diameter, and the most distant galaxies are receding from us at speeds very close to the speed of light. Hence, the time that these galaxies have been moving away from each other is approximately ten billion years. This indicates that the universe must be at least ten billion years old.

In deriving this estimate we have ignored the effects of gravity. All matter is subject to this force. For example, the motion of the planets around the sun is determined by the gravitational attraction between these objects. Likewise, gravity binds stars within a galaxy. It also operates between the galaxies themselves. The expansion of the universe is clearly related to the force of gravity. This relationship will be the subject of the next chapter.

4

Space, Time and Gravity

The theory of gravity that describes the large-scale dynamics of the universe was developed by Albert Einstein during the first two decades of the twentieth century. It is referred to as the *general theory of relativity*. It will be helpful if we now consider some of the ideas behind this theory.

We should begin by considering the speed of light. In 1865 the English physicist James Clerk Maxwell derived equations proving that electromagnetic radiation travels in a vacuum at a constant and finite speed. One of the key assumptions that Einstein later made was to suppose that two observers who are moving at a constant speed relative to one another would measure the same value for the speed of light.

Einstein's assumption goes against our intuition, to say the least. What might we expect? Speed is a relative quantity; we can measure the speed of an object only in terms of its relationship to something else. For example, when we say that a train travels through a station at a constant speed of one hundred kilometres an hour, what we really mean is that the distance between the train and the platform changes at this rate.

Let us consider two trains, *A* and *B*, that simultaneously travel through the station at this speed. If the two trains are moving in the same direction, they will appear to be at rest relative to each other. If they move in opposite directions, the apparent velocity of train *B* as measured by the driver of train *A* would be two hundred kilometres per hour.

We might imagine that velocities will always add up in this fashion. Rather surprisingly, this is not so for light. If the driver of train *A* were to send a pulse of light to the driver of train *B*, both would measure the speed of light to be the same, even though the pulse would be moving away from *A* and towards *B*. Indeed, both drivers measure the speed of light to be the same as if they were at rest!

Einstein realized that this paradox can be resolved if we are prepared to revise our ideas about the nature of space and time. Space and time are quantities that we measure. We usually measure distance with a ruler, and it is natural for us to think of time as the ticks on a clock. Yet another, equally valid, way of measuring time is to set up a lamp that emits a flash of light at regular, well-defined intervals. A unit of time can then be defined as the interval between two successive flashes. The total time elapsed between two events is then given by the total number of flashes emitted.

Consider the scenario depicted in Figure 4.1. Two observers, whom we refer to as *A* and *B*, are shown together with a moving train. Observer *A* travels inside the train, whilst observer *B* stands on the platform. *A* sets up a clock by placing a lamp on the floor of the coach and a mirror on the ceiling directly above the lamp. *A* defines a unit of time as the interval taken for a pulse of light to travel from the lamp, up to the mirror and back again.

Figure 4.1a depicts what is seen by observer *A*. Observer *A* watches the light emitted by the lamp and sees it travel straight up and then down, since *A* is at rest relative to the lamp. Figure 4.1b represents what *B* observes as the train rushes through the station from right to left. Observer *B* watches the same pulse of light, but the light appears to *B* to be moving at an angle to the horizontal. According to *B*, the light also moves sideways as it travels to the top of the coach and back again because the train itself is moving.

When the two observers meet afterwards, they disagree on the distance the pulse of light must travel to complete its round trip. Since the unit of time was defined to be the interval it took the light to return to the lamp, they effectively *measure time differently*. The difference is very small for the everyday velocities that we are used to, and we do not notice it directly. It becomes significant only when the relative speed is close to the speed of light.

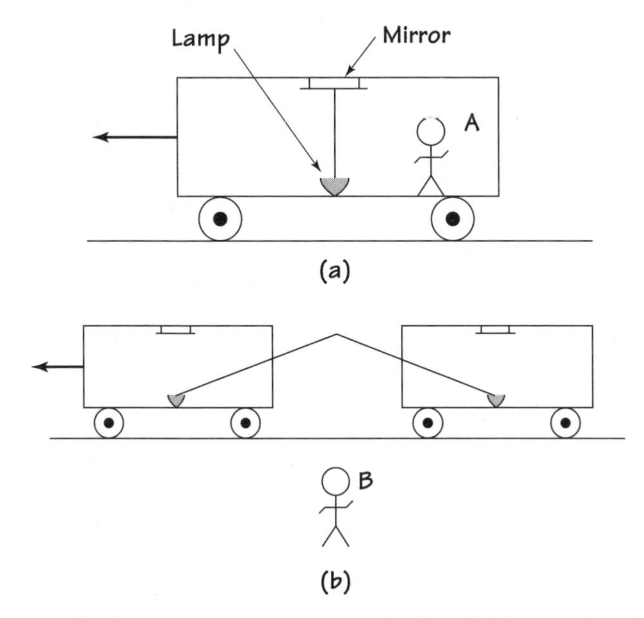

Figure 4.1. Both observers, *A* and *B*, measure the time it takes a pulse of light to travel from the lamp to the mirror and back again inside the moving train. The path followed by the light is represented by the continuous line. (*a*) *A* is travelling in the train and observes the light to move vertically upwards and downwards. (*b*) The stationary observer *B* sees the light move diagonally as the train travels through the station from right to left.

The key conclusion to be drawn is that time is intrinsically related to the motion of the person who is measuring it. This has important consequences for what we mean by the concept of time. Having come this far, though, we are faced with a new conceptual problem. If observers *A* and *B* disagree on how far the light pulse travels, which of them makes the correct measurement?

The answer is that both of them do! This apparent inconsistency is resolved in the following way. Let us take a leap of faith and consider what would happen if space and time were treated as similar quantities. Clearly space and time are not identical. We can travel both forwards and backwards in space, but we only know how to travel forwards in time. We do not know how to travel back in time and meet with our ancestors.

Despite these differences, let us treat time as some sort of extra-space dimension. We are all familiar with the three space dimensions of length, width and height. We can move up and down, forwards and backwards, and from left to right. We say that space is three-dimensional.

Combining space and time leads us to the concept of a four-dimensional quantity that is known as *space-time*.

We can easily picture in our minds what a three-dimensional object, such as a brick, looks like. However, our brains cannot cope with four-dimensional objects, so try not to picture what this four-dimensional space-time might look like. What we can do is consider a special case in which two of the space dimensions are ignored. In this simplification, there is a single dimension of space and one dimension of time, and space-time becomes two-dimensional, just like a sheet of paper. This means that we can draw a picture of this two-dimensional space-time.

Let us return to the two observers, *A* and *B*. Suppose that the high-speed train that *A* travels on completes a journey between two stations, *P* and *Q*. In Figure 4.2 we have drawn the space-time diagram for this journey. Time runs up the vertical axis, and space is represented by the horizontal axis. The movements of *A* and *B* are also shown. A stationary observer in space still moves in the time direction, so *B*, who remains at station *P*, has a trajectory denoted by the vertical line. Observer *A* travels from *P* to *Q*, so *A*'s path in space-time is given by the diagonal line. The total distance travelled by *A* in *space-time* is given by the length of this diagonal line.

Remarkably, the two observers measure the same length for this diagonal line. This is a truly unexpected result. It implies that the separation between points in space-time is *independent* of the observer, just as is the speed of light. In conclusion, the two observers agree on distances and times when they do not think of space and time as separate entities, but combine them into space-time.

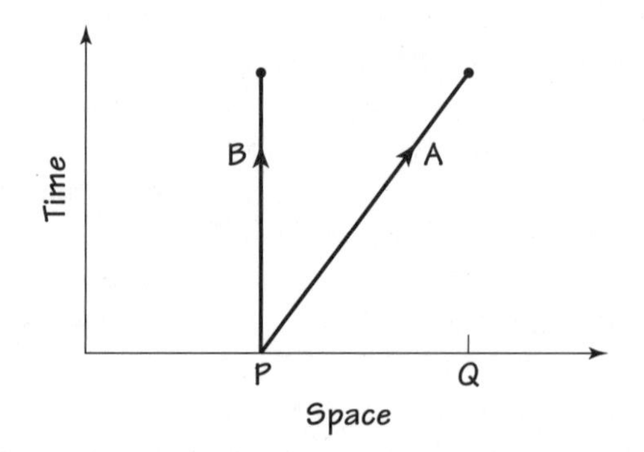

Figure 4.2. A space-time diagram illustrating the journey made by *A* between the stations located at *P* and *Q*. Observer *B* remains at *P* throughout the course of *A*'s journey.

What does this example tell us about the way we make measurements of space and time with our familiar rulers and clocks? Why did *A* and *B* measure time differently? To answer such questions it is helpful to consider an analogy. Suppose we take a brick that is shaped as a perfect cube and suspend it from the ceiling. If we then shine a light on the brick, a shadow will appear on the floor.

The surface of the floor is two-dimensional. This means that the shadow is really a two-dimensional projection of the three-dimensional brick. Even though the brick is a perfect cube with all sides of equal length, the shadow will not necessarily resemble a perfect square. It only does so if the lamp is positioned exactly above the brick. The actual shape of the shadow is determined by the relative positioning of the brick and the lamp. The shadow changes shape if the lamp is moved to a different position.

Suppose we arrange for two people to go separately into the room in which we have hung the brick. We give each a lamp and ask each person to position it somewhere above the brick. We request that each individual draw a sketch of the resulting shadow. We keep these two people separated so that neither knows what the other is doing. Each person will position the lamp differently and will draw a different picture. If we ask these individuals to measure the sides of the image that they see, they will not agree on the lengths obtained even though they are measuring the same fundamental object. They would be able to reconstruct the shape of the three-dimensional brick to be a cube only after they had accounted for the projection of the shadow on the floor.

A similar projection effect occurs when we measure space and time individually. The four dimensions of space-time are projected onto the three dimensions of space and the one dimension of time in the same way that the three dimensions associated with the brick are projected onto the two dimensions of the floor. The observers *A* and *B* measure space and time differently for the same reasons that our two observers in the room see different shadows. These differences are a direct consequence of the projection. In each case, both observers are essentially looking at the same quantity, but from a different perspective. In the brick analogy, the difference arises because the position of the lamp is altered. In the example of Figure 4.1, the difference is caused by one observer moving whilst the other is stationary.

Unification of space and time into space-time enabled Einstein to move one step closer towards his theory of gravity. We know that gravity acts on all massive bodies, including trains and people. We should therefore include gravity in the framework if we are going to arrive at a more complete picture of the universe.

We can model the effects of gravity in the universe with a sheet of polythene. Consider a hypothetical observer such as an ant that moves about on the surface of such a sheet. The ant can move forwards or backwards and also to its left or its right, but it cannot move up or down. As far as the ant is concerned, the universe consists of the two-dimensional surface of the polythene. This surface represents the 'space' of the ant's universe.

Einstein's insight was to realize that space-time becomes warped when a massive object such as a star is present. Since all massive objects are affected by gravity, this means that the force of gravity can be described in terms of the distortion of space-time.

For example, what would happen if we were to place a heavy ball onto the sheet bearing the ant? The polythene would become distorted, and the space would no longer appear flat. Instead, it would be *curved*. If the ant were to roll a lighter ball onto the polythene in the vicinity of the heavier ball, it would see the former begin to move towards the latter. This is precisely what happens when the force of gravity operates between heavier and lighter objects. Because of the gravitational influence of the Earth, when we throw a ball up into the air, the ball does not move in a straight line. Rather, it changes direction and falls back down to the ground.

The idea, then, is that we can think of gravity in one of two ways. The force of gravity due to the heavier ball acts on the approaching lighter ball. This causes the lighter ball to be attracted towards the heavier one. Alternatively, the presence of the heavy ball causes the polythene to become curved, and this curvature alters the path of the lighter ball. Either way, the lighter ball moves towards the heavier one. The amount of matter in a particular region determines to what extent the space is distorted. A larger mass results in more distortion so that the deflection of lighter objects from a straight-line path is more pronounced and the strength of gravity is stronger, as expected. In short, *the force of gravity is equivalent to a curving of the space-time.*

Describing gravity in this way is rather like trying to putt a golf ball. If the green were precisely flat, the task would be easy since we would simply have to knock the ball directly towards the hole. However, the bumps and dips typically present on the green suggest that the correct path for the golf ball is not necessarily a straight line. Instead, the golf ball will follow a curved trajectory as it falls into the dips and moves over or around the bumps. Such effects must be accounted for when aiming the putter. Similarly, we must account for the distortions in space-time that arise due to massive objects.

It should be noted that it is the complete space-time, and not just space, that is curved by massive objects, but we do not need to get involved in the subtleties of general relativity here. The key point is that the curvature provides a consistent description of the force of gravity.

Einstein's theory allows us to follow the evolution of the entire universe according to a well-defined law. We can employ his idea to investigate what the universe might have looked like when it was much younger than it is today. The theory also provides us with a more accurate picture of what an expanding universe actually looks like. According to Einstein, the galaxies appear to be moving away from each other because it is *the space between them that is expanding.*

To illustrate this concept more fully, consider the following analogy with a balloon. We take a deflated balloon and draw dots on the elastic. These dots are supposed to represent galaxies, and the elastic of the balloon represents space. The surface of the balloon can be viewed as the universe in some sense. We then place an ant onto one of these dots. As we blow air into the balloon, the elastic 'space' is stretched, and the balloon expands. What does the ant observe as this expansion of space proceeds? It thinks that its galaxy is stationary and that all the other galaxies are moving away from it.

The extension of this analogy to the real universe is straightforward. All the distant galaxies appear to be moving away from ours, and this is precisely what the ant sees. The separation in the ant's universe is due to the stretching of the balloon's surface. By analogy, it is more accurate to interpret the motion of the galaxies in our universe as being due to the expansion of space itself.

At earlier times the amount of space between the 'dots' or galaxies must have been much smaller than it is today. If we consider very early times, the amount of available space could have been so small that it literally vanished. In this simplified view, there would have been some point in time when there was no space at all. In this sense, space would be finite, since it would not have existed for an infinite amount of time.

But what about time? We discussed earlier how space and time should always be considered together. Indeed, we cannot really have one without the other. If space is finite, is the same necessarily true of time? In other words, does the universe have a finite age? The answer to this question might be yes. In this case, the complete four-dimensional space-time, and not just three-dimensional space, would vanish at some finite point in the past. In other words, time would also cease to exist beyond this point, and the universe would have a finite age.

Such a point is referred to as a *singularity* by cosmologists. It corresponds to the 'big bang' when the entire universe – including space and time – came into existence. In a very real sense, the singularity represents the 'edge' of the universe.

The expansion of a universe that begins with a big bang is shown schematically in Figure 4.3a. There is an initial singularity, and the

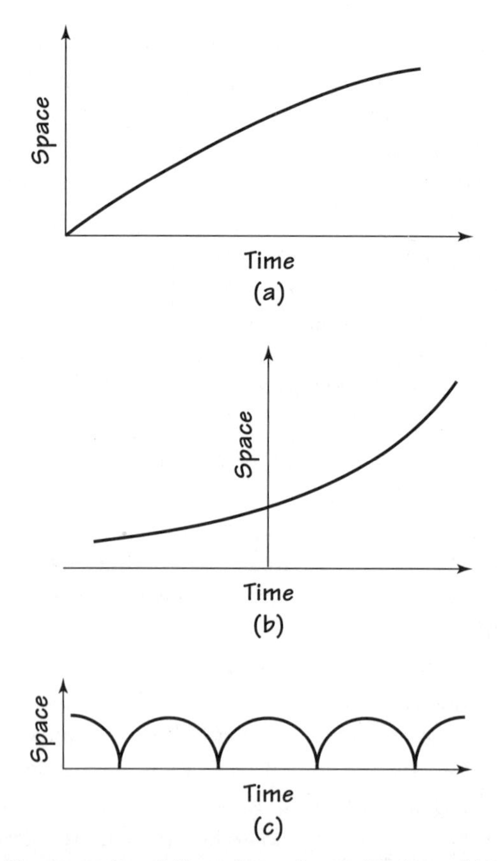

Figure 4.3. (*a*) The expansion of the universe from a big bang. (*b*) The expansion of the universe at the present era does not necessarily imply that the universe had an origin. The expansion may be such that the volume of the universe vanishes only in the infinite past. (*c*) Alternatively, the universe may undergo a perpetual series of oscillations between successive expanding and contracting phases.

expansion slows down due to the attractive nature of gravity. In some sense, the singularity may be identified with the origin of the universe, because the history of the universe cannot be traced beyond this point.

We cannot state conclusively at this stage whether Figure 4.3a does indeed correspond to our universe. Although the expansion of the universe is consistent with the notion of an origin, it does not by itself provide sufficient proof for a beginning. For example, it is possible that the universe has been expanding for an infinite amount of time, as shown in Figure 4.3b. In this scenario, the expansion is such that the universe has a finite size today, but does not have a definite origin.

A third possibility should also be considered. Our observations indicate that the universe is expanding at the present era, but this does not necessarily imply that the universe has always been expanding. Indeed, the universe may well have undergone a contraction at some point in its distant past before bouncing into an expanding phase.

This possibility suggests that the universe may be able to oscillate between expanding and contracting phases. In principle, there is no limit to the number of times that this could happen, so the cycle could be repeated indefinitely. The picture that arises is of a universe that is forever contracting and expanding, as shown in Figure 4.3c. As in the second example, this universe need not have a definite origin.

In conclusion, we have seen in this chapter that Einstein's theory of gravity leads us to the notion of an expanding universe. The expansion occurs because space-time is being stretched. The question that arises is which of the three possibilities that we outlined in Figure 4.3 corresponds to our own universe. In the following chapters, we will discuss further observational and theoretical evidence that strongly favours the scenario depicted in Figure 4.3a.

Before considering this evidence, we need to understand how matter would have behaved when the volume of the universe was smaller. The temperature of the young universe would have been so high that atomic nuclei would have been unable to survive. The matter in the universe would have consisted entirely of elementary particles. In the next chapter, therefore, we will switch from considering the largest objects in the universe – the galaxies – to the smallest objects that exist – the elementary particles.

5

Particles and Forces

Exploring the subject of elementary particles is rather like trying to find our way around an enormous zoo without the help of a guidebook to identify the different species of animal. How are we to make sense of it all? We will begin by summarizing some of the properties exhibited by the elementary particles. It is helpful to picture each particle as a tiny sphere that has three fundamental characteristics: electric charge, mass and spin. The different particles can be described in terms of these three basic quantities.

Electric charge is a familiar concept. Some particles carry it, but others do not. Those that do not are said to be electrically neutral. Likewise some particles have a mass, but others are massless. The mass of a particle contributes to its total energy, because mass is just another form of energy. Mass may be converted into energy and vice versa, and, indeed, a huge amount of energy may be produced from a relatively small mass.

The amount of spin that a particle carries determines its rate of rotation. We can view spinning particles as rotating about an axis. The electron is an example of a spinning particle. The spin of all elementary particles is severely restricted. Those particles that do not rotate have zero spin. Particles that do rotate have a spin that is directly related to that of the electron.

Elementary particles are divided into two main groups depending on the amount of spin that they carry. These groups are referred to as *bosons* and *fermions*. Their names honour the famous theoretical physicists Satyendra Bose and Enrico Fermi. Particles that have zero spin or twice the spin of the electron are examples of bosons. Particles that carry the same spin as the electron, or three times that amount, are fermions. The electron is an example of a fermion.

Boson particles behave much differently from fermions. The most important difference manifests itself when a large number of particles are confined to a very small region of space. The fermions can be viewed as hard spheres, such as marbles. There is a limit to how many fermions can be packed together in the same place, just as there is a limit to how many marbles we can place in a box. In contrast, there is no limit to how many bosons can be squashed together. We can picture bosons as being very malleable particles. In short, bosons are soft, and fermions are hard.

Twelve fundamental particles constitute the building blocks of all the matter that we observe today. All twelve have the same intrinsic spin and are examples of fermions. They may be classified into two families of six particles each. These families are known as *quarks* and *leptons*. The electron, which is a lepton, is the lightest member of its family to carry an electric charge. The two lightest quarks are known as the 'up' and 'down' quarks, respectively.

The quarks and leptons behave very differently from each other. In particular, the leptons can exist on their own, whereas the quarks are quite sociable. We never see quarks as separate entities; they always group together in twos and threes. (We will see why this is so shortly.) The proton and neutron are both examples of quark triplets. The proton is composed of two up quarks and one down quark, whilst the neutron consists of one up and two down quarks. The differences between the proton and neutron can be traced to the different properties of the up and down quarks. Thus the neutron has a slightly larger mass than does the proton because the down quark is slightly more massive than the up quark.

Why is it that the quarks must be bound together into pairs and triplets, whereas the leptons may remain free? To answer this question, we need to establish how the particles interact with one another. In other words, we must determine the forces that act between them. In the previous chapter, we discussed the force of gravity, which influences all elementary particles. However, there are three other basic forces besides gravity. These are the electromagnetic force, the strong force and the weak force. We have already seen how gravity can be described in terms of the curving of space-time. How can we describe these other three forces?

Let us begin with the force of electromagnetism, which is the force that operates between electrically charged particles. Opposite electric charges attract, whereas like electric charges repel. Consider what happens when two electrons – with equal negative charge – travel towards each other from opposite directions. As they approach, they start to slow down, and, at a certain point, they then accelerate away from one another. The force of electromagnetism has changed both the speed and the direction of motion of the two particles.

On one level, this process is rather like that of kicking two soccer balls at each other. When they collide head-on, the balls bounce back in opposite directions. The analogy is not quite perfect, because unlike the soccer balls the electrons do not actually touch each other. Instead, as they approach each other the two particles exchange information. At this point they discover that they have the same electric charge and therefore must move away from each other.

How is this information carried between the two electrons? To answer this question we must make a small digression. Consider the inside of a closed container that is devoid of matter and radiation. This represents a *vacuum*. The vacuum is defined as something that is completely empty. It has zero energy. The question now arises as to how we would determine experimentally whether the inside of the box really was empty or not.

We would have to 'look inside' the container by sending in a beam of light and measuring the interior. But light travels at a finite speed, and it takes a finite amount of time to complete such a measurement. Because we can establish that the box is truly empty only once the light has reached our detectors, we have no way of establishing whether the box was empty *during* the measurement process. In reality, the energy inside the box could fluctuate slightly from zero. The only criteria are that the fluctuation not influence the light pulse in any way and that the fluctuation be over before the measurement is completed.

Thus the energy contained within the box cannot be determined precisely, even when we have access to perfect measuring equipment. There is always an intrinsic uncertainty about the amount of energy present at any given instant. Furthermore, since mass is equivalent to energy, this fluctuation in energy can manifest itself as the spontaneous creation of particles out of the vacuum. As long as these particles have decayed back to nothing before the measurement is completed, they cannot be directly detected.

Remarkably, this is what actually happens! Particles are continually being created out of nothing as energy fluctuations and then destroyed again, but they are destroyed so quickly that they can never be directly observed. Such particles are referred to as *virtual* particles.

This process is rather like watching the classic comedy routine where a villain is hiding from the police. The police officer has a feeling of being watched, but when he looks round, he sees no one. As soon as the officer's back is turned, the villain jumps out from behind a tree and pulls a face at him. By the time the officer has turned round again, the villain is once more behind the tree. In this analogy, the police officer represents the experimentalist, who makes a 'measurement' by turning around. The undetectable villain represents the virtual particle.

Virtual particles must always appear and disappear in pairs. To understand why, consider an electron that pops into existence out of nothing. Before the electron appears, the effective electric charge is zero since no particles are present. The electron carries a negative electric charge. Since the total electric charge is always conserved in any physical process, the electron must be accompanied by a second particle that carries an equal amount of positive charge. This ensures that the charge of the electron is exactly cancelled by the opposite charge of the second particle. The overall charge remains zero throughout.

The second particle in this pair of virtual particles is known as the *positron*. The positron can be viewed as the *antiparticle* of the electron. The positron can also exist in nature as a real particle in the same way that the electron does, but it cannot survive for long in the present universe. In general, all types of particles, such as the quarks and leptons, have antiparticles associated with them. A particle and its antiparticle have the same mass and fundamental spin, but the two have *opposite* electric charges. (If the particle is electrically neutral, the antiparticle will be as well.)

Particles do not get on well with their antiparticles. They rapidly annihilate each other whenever they meet. They can be pictured as two opponents who are fighting a duel with pistols. Rather than standing twenty paces apart, the adversaries stay very close to each other, and the outcome is that both are hit when they fire their pistols.

In short, a virtual particle created out of nothing is necessarily accompanied by its antiparticle. The inherent antagonism between particles and their antiparticles ensures that they annihilate each other very soon after their creation. Virtual particles have such short lifespans that we never have sufficient time to observe them directly. This is why they are called virtual particles. The key feature of all virtual particles is that they are created literally *out of nothing*.

We have a problem, however. The process we have just described seems to violate the conservation of energy. If energy is always conserved, the total amount beforehand must be the same as the amount afterwards. In other words, we cannot create energy out of nothing. The idea behind this principle is neatly summarized in the familiar phrase: 'there is no such thing as a free lunch'.

The vacuum must appear empty when we observe it and must have zero energy associated with it. The problem, then, is this: if there is no energy available beforehand, what is the source of energy that allows virtual particles to be created. Are we not getting something for nothing here?

Well, not necessarily. Consider what happens to electric charge when an electron and its antiparticle are created. The total charge is zero throughout, but for a brief moment this total becomes separated into equal amounts of positive and negative charge. This process is equivalent to adding plus one and minus one; the overall result is still zero.

A similar argument applies to the energy of the particles. The electron has positive energy, and, in a certain sense, the antiparticle has an equal amount of *negative* energy. Thus the total energy remains zero throughout the creation and subsequent annihilation of the virtual particle and its antiparticle.

Because we are used to energy being positive in our everyday experience, this concept of negative energy may seem a little strange initially, but some forms of energy can indeed be negative. Negative energy is particularly important for understanding how the universe might have come into existence in the first place. The concept will also play a significant role in our discussion of black holes.

One example of negative energy is the energy associated with the force of gravity. Consider an object that is located at the bottom of a hill. We must work to overcome the force of gravity in order to lift this object and carry it up the hill. We lose energy in the process of lifting and carrying the object. Since the total amount of energy present must always be conserved, the object must gain the energy that we lose. The object must have more energy when it has been brought to the top of the hill than it had at the bottom.

We can reverse this argument and start with the object at the top of the hill. If we were to throw this object down the hill, the object would lose energy, since it has to have less energy when it is at the bottom. Losing positive energy in this way is equivalent to gaining *negative* energy.

What we have seen, then, is that the vacuum is not the simple place that we thought it was. It *appears* to be empty when we measure it, but the reality is quite different. In fact, the vacuum is a very busy environment with virtual particles continually being created and destroyed.

In general, fluctuations in energy are called *quantum fluctuations.* The inherent uncertainty in measuring the energy of the vacuum applies to all physical systems, including real elementary particles. This observation implies that the energy of a particle can never be measured precisely. But if we cannot determine its energy, we can never determine exactly either how fast the particle is moving or where it is located. This ambiguity in the position of a particle due to quantum fluctuations will prove central in later chapters when we discuss the origin of galaxies and, ultimately, of the universe itself.

We are now in a position to return to our original question of how the electrons are able to communicate with each other and discover that they have the same electric charge. The answer will enable us to describe how the forces of nature are transmitted.

The general idea is that two real particles exchange virtual particles continually. The role of these virtual particles is to carry information between the real particles. It is the virtual particles that tell the real particles that they are about to meet. Depending on the nature of the information they receive, the real particles then change direction. As observers of this process, we interpret such a change in direction, as well as in velocity, as being caused by a force. This force arises due to the exchange of virtual particles.

When two electrons meet in this way they exchange what are known as virtual *photons*. This exchange is shown in Figure 5.1. The paths followed by the electrons are shown by the solid lines, and the virtual particle is drawn as a wavy line. The photon tells the electrons that they both have the same electric charge and should move away from each other.

A photon, which has zero mass and electric charge, may be viewed as the particle equivalent of electromagnetic radiation. Since the energy

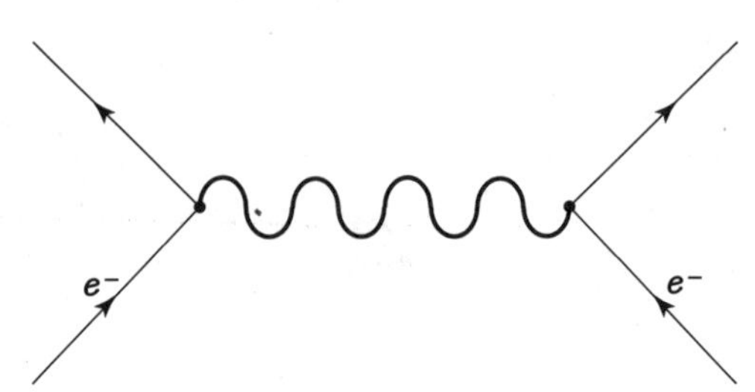

Figure 5.1. Two approaching electrons e^- repel each other by exchanging a virtual photon, as shown by the wavy line.

of an electromagnetic wave increases as its wavelength decreases, a very high energy wave, such as a gamma ray, has a very short wavelength. In this example, it is often more convenient to view the wave as a particle positioned where the crest of the wave would be.

It helps to view the approaching electrons in Figure 5.1 as two skaters. The skaters move towards one another, and once they are sufficiently close, each throws a brick towards the other. These bricks represent the virtual photons. The skaters must change direction when they exchange bricks in this way to conserve momentum. In effect, they must move in the opposite direction to that followed by their newly released bricks. They therefore move away from each other, and this is precisely what happens to the electrons.

Thus far we have discussed how electrons are repelled from one another by virtual photons. This description of force in terms of virtual-particle exchange is not confined solely to electrons. We may employ the idea to describe the other forces as well. The generic picture is that each force in nature is carried by a particular set of virtual particles. The defining feature of each force is then determined by the properties of the virtual particles associated with it. Each force is different because each is transmitted by a different type of virtual particle. When any two particles in the universe meet, they exchange various virtual particles, and the precise choice of exchange particles is determined by the properties of the real particles in question. It will be useful at this stage to summarize the types of forces that operate in the universe today:

1. *The electromagnetic force:* The electromagnetic force acts between *all* particles that carry electric charge. Electrons, quarks and protons are subject to this force, but neutrons are not. Inside atoms, this force holds the negatively charged electrons in orbit around the positively charged nuclei. The virtual particles associated with the electromagnetic force are photons.

2. *The weak force:* The weak force influences all quarks and leptons and is responsible for some kinds of radioactive decay. As its name suggests, this force is rather weak. We do not experience it directly as we do electromagnetism and gravity because the weak force has a very short range. It can operate only over distances that are one hundred times smaller than the size of a proton. Two particles must be at least that close if they are to interact via the weak force. If they are separated by any greater distance, the particles will not notice the effects of the weak force.

Why is the range of the weak force so small? In general, the range of a force is determined by the masses of the virtual particles that are exchanged. Suppose we were to view a virtual particle as a single wave. We have seen previously how a photon may be viewed as the particle analogue of an electromagnetic wave. In a similar way, particles that have a mass also have wavelike properties. Now, we have also seen that the energy of a wave is related to its wavelength. A higher energy corresponds to a shorter wavelength. Since energy is equivalent to mass, a higher mass also corresponds to a higher energy and therefore to a smaller wavelength.

In this sense, the range over which a massive virtual particle can influence the real particles is determined by its wavelength. A higher mass results in a shorter range and vice versa. The weak force is carried by the exchange of virtual particles known as the *W* and *Z* particles. These particles are very massive compared to particles such as the electron and proton. It is this feature that results in the short range of the weak force.

3. *The strong force:* We have already mentioned that the nucleus of an atom consists of protons and neutrons that are tightly packed together. This is slightly enigmatic because we know that protons have positive electric charge. We might expect protons to be repelled from one another, especially as they are so close together in the nucleus. From what we have seen thus far, the protons should be interacting through the electromagnetic force by exchanging virtual photons. This exchange should cause the nucleus to explode violently. What is it then that stops the nucleus from disintegrating?

The protons in the atom's nucleus do indeed interact via electromagnetism, but another force also operates between the protons and neutrons. This force is much stronger than the electromagnetic force, so the latter's influence becomes negligible inside the nucleus. This additional force is an attractive force, and it keeps the nucleus together.

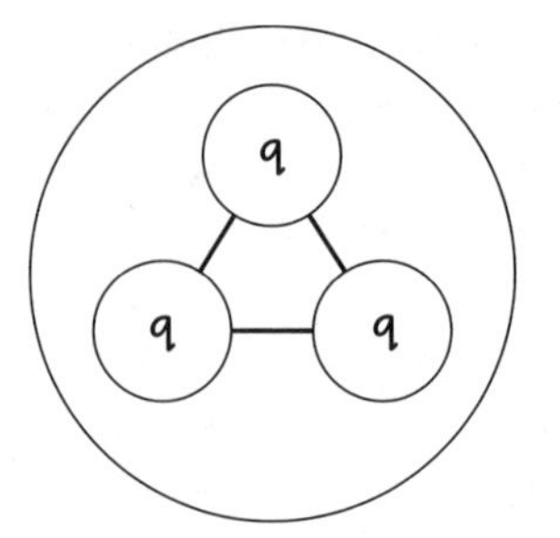

Figure 5.2. The proton is made up of three quarks, q. The quarks are bound together by gluons. These are represented by the straight lines and may be viewed as tiny elastic bands.

This force, known as the strong force, is transmitted by particles called *gluons*. These particles are so called because they effectively 'glue' the nucleus together.

In fact, the strong force operates between quark particles. It is responsible for keeping the quarks locked together inside the neutrons and protons. Leptons do not feel the effects of the strong force, and this is one of the key reasons why quarks and leptons are so different.

Let us consider how the three quarks inside a proton interact with one another. The internal structure of the proton is shown schematically in Figure 5.2. Each quark is bound to the others by gluons. In Figure 5.2 these gluons are represented by the straight lines.

The gluons may be thought of as tiny elastic bands that join the quarks together. It is this feature that results in some of the unique properties of the strong force. When the quarks are close to one another inside the proton, the elastic bands are unstretched and relatively unimportant. Hence, the quarks are effectively free and do not notice the presence of the gluons. But if the quarks attempt to move away from one another, the elastic bands become stretched and pull the quarks back together. The strength of the strong force therefore *increases* when the quarks are farther apart. This is in contrast to the other three forces of nature, whose strengths decrease as the two interacting particles separate.

The elasticity of the gluons prevents the quarks from separating by more than about 10^{-15} metres. The quarks are simply pulled back into the proton if they attempt to escape from it. Once quarks become bound, they cannot escape from one another.

This is not the full story, though. Naively, one might think that a huge attractive force should exist between two quarks that are separated by very great distances. We may consider as an example two quarks that are contained within two different protons. If the strong force is acting between these quarks in the same way that it acts between quarks within an individual proton, the two protons should be very strongly attracted to each other. But if this were the case, all the neutrons and protons in the universe should be attracted to one another in a similar way. The end result would be a giant cosmic nucleus!

This does not happen because bound quarks tend to interact only with their immediate neighbours. It is helpful to think of each quark in a proton as having just one elastic band associated with it. Once a quark has been joined to another quark in the manner shown in Figure 5.2, the particles find it very difficult to interact with other quarks in other protons. Thus, the range of the strong force between different protons is very short. Two protons separated by more than 10^{-15} metres will not feel its effects. However, when protons are closer than this critical distance, they are attracted to one another. Protons and neutrons inside an atomic nucleus are about this close, which is why they remain bound together.

4. *Gravity:* We saw in the preceding chapter how the force of gravity can be described in terms of the curvature of space-time. Is there an alternative description of gravity that can be given in terms of the exchange of virtual particles? Since the other three forces of nature can be modelled in this fashion, such a description might be possible. A great amount of time has been spent by many researchers in an attempt to develop such a theory of gravity. In the picture that has emerged from this effort, the force of gravity is carried by a virtual particle known as the *graviton*. However, a consistent mathematical description of such a process that is valid in the limit where the gravitational interactions are very strong has yet to be developed, and whether such a description is possible in this limit is still not clear.

To conclude, we have seen in this chapter that the fundamental particles may be classified into two groups known as the quarks and the leptons, respectively. Quarks and leptons behave quite differently. In particular, the leptons can exist as free particles because they are not influenced by the strong force. The quarks are affected by this interaction and must group into twos and threes. The protons and neutrons that make up the nuclei of atoms are examples of quark triplets.

The electromagnetic, weak and strong forces are all carried by the exchange of virtual particles that are created out of the vacuum. These forces have different properties because they are associated with different virtual particles. For example, the range of the electromagnetic force is unlimited, but the weak force has a very short range. This difference arises because the particle carrying the electromagnetic force – the photon – is massless, whereas the *W* and *Z* particles of the weak interaction are massive.

Presently, the electromagnetic force is about one hundred times weaker than the strong interaction, and the weak force is about one hundred thousand times weaker. Although significantly weaker than the other three forces, gravity has infinite range and can therefore act over cosmological distances. Also, gravity affects all particles, whereas electromagnetism can influence only charged particles. This property of gravity, together with the fact that the universe is electrically neutral on large scales, is why in the present era it is gravity, and not the other forces of nature, that determines the overall dynamics of the universe.

The relative strengths of the forces depend quite strongly on the temperatures involved. This observation is perhaps not too surprising. We know, for example, that the properties of a substance can change as the temperature is increased. A block of ice will soon melt and then evaporate into steam if it is placed inside a hot oven. In a similar way, the virtual particles that carry the forces also behave differently at very high temperatures. This fact is important for our study of the early universe. In the next chapter we will investigate how the four forces of nature are predicted to behave at very high temperatures and energies.

6

Grand Unification, Higher Dimensions and Superstrings

The energy scale of processes that occur naturally on the surface of the Earth is relatively low. At these energies, the forces of nature are different from one another because the particles that transmit them have very different properties. However, above 10^{15} degrees the electromagnetic and weak forces become indistinguishable. That is, they take on similar identities. Indeed, such modifications have been observed in the large particle accelerators. This fundamental similarity between the electromagnetic and weak forces suggests that they might be two components of a more fundamental force. The two forces are said to become *unified* into the *electroweak force* above 10^{15} degrees.

What causes such a unification at this particular temperature? We know that temperature is just a measure of the amount of energy present; higher temperatures correspond to higher energies. Furthermore, energy and mass are equivalent. In this sense, then, mass is closely related to temperature. For every mass there is a corresponding energy and temperature.

The masses of the W and Z particles correspond to this temperature of 10^{15} degrees. These particles have two types of energy associated with them: one is the energy stored in their mass, and the other is the energy determined by how fast they are moving. This latter type is called *kinetic energy*. As the temperature increases, so does the kinetic energy of the particles. The energy stored in the mass remains fixed, however, because the masses of the particles are constant.

At very high temperatures, the kinetic energy of the particles dominates the energy contained within their mass. As a result, the particles will behave as if they had no mass at all! For the *W* and *Z* particles, the kinetic energy begins to dominate at 10^{15} degrees. Thus these particles *behave as if they were massless particles* above this temperature, just like the photon.

The unification of electromagnetism with the weak force has an important consequence. It means that there will be only three effective forces in operation above 10^{15} degrees. These will be the electroweak, strong and gravitational forces. Because electromagnetism and the weak force do become unified, it is natural to ask whether the strong and electroweak forces combine at an even higher temperature. Can we find out if this is the case?

It should be emphasized that 10^{15} degrees represents the highest energy scale that can be probed directly by current technology. Once we move above this temperature, we enter the realm of theoretical speculation. Having said this, there is strong theoretical evidence to suggest that the strong and electroweak forces do indeed become unified when the temperature reaches 10^{27} degrees. The two forces combine under what is collectively termed the 'Grand Unified Theory', often abbreviated to the 'GUT'.

The reasons for this further unification are essentially the same as those we discussed earlier in the electroweak case. The gluons, photons and *W* and *Z* particles all start behaving in a similar fashion, so the forces become indistinguishable. The Grand Unified theory also predicts that new exchange particles – called *X* particles – should become important at this temperature. *X* particles cause quarks to decay into leptons and vice versa. They have a mass that corresponds to a temperature of 10^{27} degrees, and they decay very quickly. The effects of the *X* particles are negligible below this temperature because they decay before they have had a chance to travel between two interacting particles. At higher temperatures the two interacting particles are able to get closer to one another, and at 10^{27} degrees, they become so close that the exchange of *X* particles is possible.

The precise form of the GUT of particle physics is not known at present. The high temperatures at which it becomes possible to observe the phenomenon of grand unification are beyond the reach of the most powerful particle accelerators in the world. But this has not stopped theoreticians from speculating as to what such a force might look like! We do know that it must contain the strong, weak and electromagnetic forces as components, and this requirement constrains the number of possible candidates.

One testing ground in which the predictions of these Grand Unified theories may be investigated is found in the very hot conditions that were prevalent in the big bang. At sufficiently early times the temperatures would have been above the temperature at which the GUT force takes over. If the unification idea is correct, this force should have dominated the universe at these early times. Moreover, it would have had a significant effect on how the universe came to be in its present state.

This suggests that we might be able to probe the GUT by combining particle physics and cosmology. The idea is to begin with a particular Grand Unified theory and develop a theoretical model that traces the history of the universe from the earliest moments through to the present day. This procedure leads to theoretical predictions of what the universe should look like. In principle, these predictions can then be compared to the actual universe we observe. If observation and theory are found to be in reasonable agreement, one can argue that the Grand Unified theory might be accurate. If there is disagreement, one must conclude that the particular theory in question is flawed. Indeed, this approach has already been employed over the past decade to rule out some of the simplest GUTs.

To summarize thus far, the electromagnetic, weak and strong forces become unified at very high temperatures, suggesting that these three forces have a common origin. But what is the nature of this origin? Moreover, can gravity be brought into the picture? We have seen how gravity manifests itself as the warping of space-time, but how is this related to the GUT and virtual particles? We shall proceed to address these and related questions in the remainder of this chapter.

When Einstein's theory of general relativity was published, the existence of the strong and weak forces was unknown. Scientists thought that nature was governed by the two forces of gravity and electromagnetism. The theory of general relativity explained gravitational effects but failed to account for the electromagnetic force.

In 1919 the mathematician Theodor Kaluza wrote to Einstein and suggested how gravity and electromagnetism could be combined into a *single* force. Einstein had formulated his theory within the context of a four-dimensional space-time. This was perfectly reasonable since we observe only three dimensions of space and one dimension of time. Kaluza did something quite remarkable. He introduced an *extra space dimension* and formulated Einstein's theory within the context of four dimensions of space and a five-dimensional space-time. He then proved that the five-dimensional theory is equivalent to Einstein's four-dimensional general relativity plus the force of electromagnetism. In other words, introducing an extra dimension of space into the universe is equivalent to introducing the force of electromagnetism. This implies that gravity in five dimensions equals gravity plus electromagnetism in four dimensions!

Although Kaluza's theory was acceptable from a mathematical point of view, there was a serious problem with it. Kaluza failed to explain why the fifth dimension is not observed in the real world. If this extra dimension really does exist, where might it be hiding? We can answer this question with the help of an analogy. Let us consider a length of hose-pipe and ask how one might proceed to construct such an object.

A hosepipe is nothing more than a hollow cylinder. The simplest way to construct it is as follows. We begin with a straight line, as shown in Figure 6.1a. We then attach a small circle to each point on this line, as shown in Figure 6.1b. The result is a tube, or hosepipe, whose cross-section is determined by the radius of the circles that were placed onto the line.

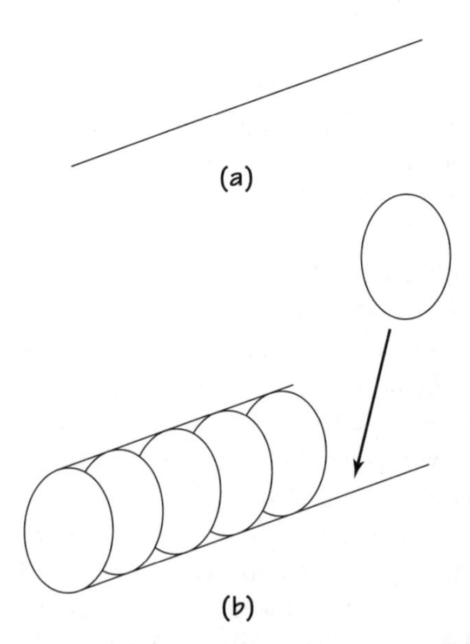

Figure 6.1. A hosepipe may be constructed by placing a small circle on each point of a straight line.

Consider what happens as we move farther away from this pipe. As the distance between us and the pipe increases, its apparent cross-section seems to get smaller relative to its length. Eventually, the apparent radius of the pipe becomes so small that our eyes cannot tell whether it has a cross-section or not. As far as our observations are concerned, the hosepipe resembles a line.

In reality, we know that the hosepipe is a cylinder. It may be viewed as a two-dimensional surface. Yet a line is only one-dimensional. What has happened as we walked away from the pipe? If we had only our observations to go on, we would think that the hosepipe was simply a line because *the second dimension associated with it is too small to be detected.* We can see only one of the dimensions because the second dimension is wrapped up into a tiny circle. Thus we would conclude from our observations that the hosepipe is a line rather than a cylinder.

The idea is that the space-time of our universe is qualitatively similar to the hosepipe. Three of the space dimensions are large and observable. There is nothing preventing us from adding a small circle to each point in space-time, in the same way that we placed a circle onto each point of the line in Figure 6.1b. When we do this, we effectively introduce a fifth dimension into the universe. We do not 'see' this fifth dimension in our everyday lives – in the sense that we are not directly aware of its existence – for the same reasons that we cannot see the extra dimension of the hosepipe when we are far away from it. The radius of the extra dimension is too small to be observed directly with laboratory instruments, but its effects are manifested in the 'big' dimensions as the force of electromagnetism.

These speculative ideas of Kaluza, although developed by the Swedish mathematician Oskar Klein in 1926, were largely ignored until quite recently. They have since become very popular. One reason for their popularity is found in the following question: if the force of electromagnetism can be understood in terms of a fifth dimension, are the strong and weak forces also related to extra spatial dimensions?

Presently, the answer to this question is unknown, although the general feeling is that the idea is very promising. We have already discussed some of the evidence that suggests that the three forces may have a common origin, but have not yet identified what the connection between them might be. If all three can be described in terms of extra dimensions, these dimensions would provide the missing link that we are looking for.

The suggestion, then, is that the universe contains extra, hidden dimensions, and these determine the characteristics of the strong, weak and electromagnetic forces. In general, the extra dimensions are referred to as the *internal space*. The internal space contains all the small dimensions that we cannot observe directly. The three spatial dimensions that we do observe are referred to as the *external space*. The structure of the external space is determined by the force of gravity, and the form of the internal space is related to the other three forces.

There are two crucial questions that must be addressed in this picture. Firstly, what is the precise structure of the internal space? Secondly, if there are extra dimensions present in the universe, why do only three of them expand to cosmological sizes?

One feature of the internal space is that it must be extremely small. Experimentalists have probed the structure of space down to scales that are many times smaller than the size of the proton. If the internal space extended over larger scales than these, it would already have been detected. No evidence for internal structure has been found, so we must conclude that the internal space is much smaller than the tiniest elementary particle.

We also have a second piece of information regarding the nature of the internal space. It cannot have any edges. If it did, particles travelling along it would eventually fall off in much the same way that marbles roll off the surface of a table. It follows that the internal space must be wrapped up in some way. That is, it must be compact and finite.

Finally, we know from our discussion in Chapter 4 that the curvature of space is related to the force of gravity. Einstein's theory works very well on large scales, so we do not want to modify it. This requirement further restricts the structure of the internal space because if the internal space was curved, additional gravitational effects would be introduced into the universe, and these would modify Einstein's theory in a significant way. We require that gravity be unaltered by the internal space. Since the absence of gravity corresponds to a flat space, the internal space should also be flat.

How can we ensure that the internal space not only is wrapped up on itself but also is spatially flat? It is helpful to consider a two-dimensional model. A piece of paper is an example of a flat surface, as is a sheet of polythene. Unfortunately, both these surfaces have edges and are not suitable as they stand. What we need to do is take our polythene sheet and modify it in such a way that the edges are eliminated, which would ensure that particles rolling along it would not fall off.

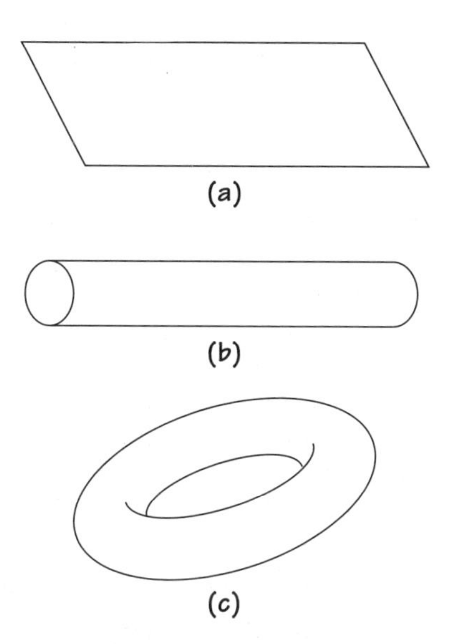

Figure 6.2. (*a*) Illustrating the local flatness of a doughnut. (*b*) A cylinder may be constructed by taking a flat sheet of polythene and joining together two of the edges that are opposite to one another. (*c*) A doughnut is then produced by bending the cylinder in such a way that its two ends meet.

Figure 6.2a shows the original polythene sheet when it is stretched out. As can be seen, there are four edges to it. We can eliminate two edges by joining two of the opposite sides together, the equivalent to rolling the sheet up into a cylinder, as shown in Figure 6.2b. There are now only two edges remaining, and these are associated with the two circles at the ends of the cylinder. These can also be joined together, as shown in Figure 6.2c. The final result resembles a doughnut.

This doughnut may not appear to be flat, but in fact it is. We began with a flat surface, and all we have done is to join up the four edges. We have not added or taken anything away from the elastic sheet that we started with in Figure 6.2a. This is important, because introducing even a small amount of matter would cause the sheet to become curved, as we saw in Chapter 4. Thus, if the surface of the elastic was flat initially, it must still be flat afterwards.

The surface of the doughnut is a good example of a finite, flat space. It has two dimensions, since it was originally constructed out of a two-dimensional polythene sheet. The actual internal space of the universe (if it exists) probably has more than two dimensions. Indeed, some of the more popular theories predict that the internal space should have *six* dimensions, making the complete universe ten-dimensional. To a first approximation, though, we may think of the internal space as being the six-dimensional equivalent of a doughnut.

In this picture, the electromagnetic, strong and weak forces arise because an internal space has been placed onto each point of the four-dimensional space-time. The precise structure and size of the internal space determines the nature of the forces that act on the particles. The fourth force – gravity – arises because the three big dimensions of space are curved.

This idea is attractive, because it provides a common origin for all the forces of nature. These forces arise because the dimensions of space have been twisted and wrapped up in an appropriate fashion. This suggests that the common origin among the forces of nature might simply be the overall shape of the higher-dimensional universe.

These theories are generally referred to as Kaluza-Klein theories. At present, no one has successfully constructed a complete Kaluza-Klein theory that correctly predicts what particles exist in the universe and how they interact. Perhaps no one ever will. These theories do seem to be a step in the right direction, though, at least from a philosophical point of view. They provide a framework for describing all of the forces of nature in terms of the geometry of the universe.

If such a link exists between the forces, it is quite plausible that the forces all have a common origin. Furthermore, it may be possible to unify them into a single 'superforce'. We have seen that there are only two effective forces in operation above 10^{27} degrees. These are the Grand Unified force and gravity. As the temperature has been steadily increased, the electromagnetic, weak and strong forces have become unified. The idea is that the GUT force and gravity should also become unified at some point when the temperature is sufficiently high.

Moreover, if the four forces are just different manifestations of the same superforce, can we describe all the particles of nature in terms of a single object as well? In other words, are the quarks and leptons just different manifestations of a more fundamental 'superparticle'?

A theory that described our universe in these terms would literally contain all of physics. In a sense, it would represent a 'theory of everything'. Such a theory could be called the 'TOE' for short.

How can we investigate what the universe's TOE might look like? We first need to establish at what point the GUT force and gravity begin to behave in a similar fashion. If the idea of higher dimensions is correct, the properties of the forces are related in part to the sizes of the dimensions. In particular, the nature of the GUT force should be related at some level to the size of the internal space. The internal space is probably as small as 10^{-35} metres. The external space, which is essentially that part of the universe affected by the force of gravity, extends for at least ten billion light years.

It is no wonder that there is such a difference between the forces! However, the external space would have been tiny during the very first moments of the universe's history. In principle, we can consider going back to a time when the external space had a size comparable to that of the internal space. At that time all the spatial dimensions in the universe would have been similar to each other. It is here that we expect some of the similarities between the GUT force and gravity to become apparent.

Considering earlier and earlier times is equivalent to raising the temperature. The internal and external spaces would have had similar sizes when the temperature was a staggering 10^{32} degrees, and this corresponds to the point where the GUT force and gravity merge into the superforce.

At this temperature the collisions between the particles would have been intense. The particles would have been able to get so close to one another that their internal structure would have become important. We must account for this structure when discussing the unification of gravity with the GUT.

An analogous complication arises when making a map. If you were constructing a map of the entire world, you would be interested primarily in the coastlines of the continents and, perhaps, the location of the world's largest mountains and major cities. A relatively small mountain would not be important to you. If you wanted a map of the region surrounding a particular mountain, however, that mountain would play a crucial role. The lay of the land would gradually become more critical as the scale was increased. Similarly, their internal structure becomes more significant as the particles get closer and closer.

Unfortunately, we cannot see 'inside' elementary particles, so we will have to make an inspired guess as to what they might look like. Many researchers take the view that the quarks, leptons and virtual particles that mediate the forces may not behave as tiny spheres after all. Instead, the preferred view is to regard the particles as extended objects that resemble pieces of string.

The theory that describes particles in terms of strings is called the *superstring theory*. According to this theory, the stringlike properties of the particles do not become evident until a temperature of 10^{32} degrees is reached.

Superstrings have elastic properties and can be thought of as minute elastic bands. Elastic bands have a tension associated with them. They become taut as we stretch them and their tension increases. It is this tension that forces an elastic band back to its original shape once it is released. The same is true for the superstring. The tension of the superstring increases as the temperature drops. This tension becomes important below 10^{32} degrees and causes the string to shrink rapidly to a point, which explains why the elementary particles we observe today behave as pointlike objects.

Strings can be either open or closed. An open string can be viewed as an elastic band that has been cut, whereas a closed string resembles a band that remains intact. Strings introduce a new feature into the picture. Their elastic nature implies that they can vibrate in the same way as guitar strings.

What are the implications of this? A wave with a shorter wavelength has a higher energy. Energy is equivalent to mass, and thus we can associate the energy of the wave with a mass. We can also associate mass with particles, so the different vibrations of the string can be interpreted in terms of different particles.

The attractive feature of this picture is that it enables us to view *all* particles in terms of the same fundamental object – the superstring. Although there are many different types of particles in the universe, the superstring theory tells us that they are all related to one another in a very fundamental way. The particles are essentially just individual superstrings. The characteristics of the superstring, such as its tension and energy of vibration, can vary, and these variations manifest themselves as particles with different properties.

Another appealing feature of the superstring theory is that particle interactions are naturally explained by pieces of string splitting apart and joining together. For example, consider a particle decaying into two lighter particles. This can be viewed as a piece of string splitting into two smaller pieces. The reverse reaction, in which two particles combine, arises when two separate pieces of string meet and 'tie the knot'.

This picture of particles as strings is very appealing. We should emphasize, however, that the superstring concept is just a theory at this stage. Although it is a very compelling idea, there is no direct observational evidence to suggest that it is necessarily correct. One of the main problems with the superstring theory is that it makes few predictions that can be tested. We do not yet know whether it leads to an accurate description of the universe.

One prediction that the superstring theory does make is that the universe should be ten-dimensional. Since we must have one dimension of time, this leaves us with nine space dimensions. Three of these must make up the world that we are familiar with, so the other six must form an internal space, and this has to be wrapped up in the way that we described in Figure 6.2. According to the theory, the size of this internal space is roughly the same size as the superstring itself, that is, 10^{-35} metres.

How does the internal space affect the vibrations of the superstring? To answer this question it will be helpful to consider an open string that has both of its ends fixed so that it behaves rather like a guitar string. When a guitar string is plucked, waves travel along it. Because the ends of the guitar string are fixed, only waves with a certain wavelength can propagate. These wavelengths are determined by the length of the string. The superstrings vibrate in a similar fashion to the guitar string because they are restricted by the internal space. In general, the wavelengths of the vibrations allowed on the superstring are determined by the structure and size of the internal space.

The superstring picture allows us to understand more fully how the structure of the internal space determines the forces acting between particles. The forces are transmitted by the exchange of the virtual particles. The properties of these forces are determined by the properties of the exchange particles. In turn, the characteristics of these exchange particles are determined by the vibrations on the superstring. These vibrations are themselves determined by the internal space. Hence, the structure of the internal space plays a crucial role in determining the nature of the forces as we observe them.

Although the idea of an internal space tells us how there can be small dimensions in principle, it does not explain why three dimensions in particular have grown to cosmological sizes. Is this number special, or is it just a coincidence? Is there some additional physical principle in operation that prevents a different number of dimensions from becoming large? The answers to these questions are largely unknown, but some suggestions have been made. We will return to such issues in Chapter 10 when we discuss what is known as the inflationary universe.

To conclude, let us recall what we said at the beginning of Chapter 5. We suggested that discussing the subject of elementary-particle physics is similar to finding our way around an enormous 'zoo' of different particles. After our journey through the world of virtual particles and higher dimensions, we have found that there may be really only one animal in this zoo. It is the very impressive specimen known as the superstring.

7

The Big Bang

We are now in a position to talk meaningfully about the earliest stages of our universe's history. Over the last few decades a scenario has emerged that indicates that the universe began life as a rapidly expanding and intensely hot fireball. This picture is referred to as the *big bang model.* Most cosmologists agree that this model represents an accurate description of the very early universe, at least for times after about one second.

In this chapter we will trace the history of the universe from the big bang through to the present day. Let us begin by discussing some of the key principles that governed the behaviour of the universe during the big bang.

The temperature of the universe at any given time is directly related to the size and age of the universe. It often proves convenient to measure the age of the universe directly in terms of its temperature. A higher temperature then corresponds to an earlier time. For example, when the universe was one second old its temperature was about ten billion degrees. It will be useful to keep this temperature in mind for comparative purposes in our forthcoming discussion.

The very early universe was considerably hotter than ten billion degrees. Matter in the form of atoms would not have been present. Indeed, atoms did not come into existence until the universe was about three hundred thousand years old. Moreover, nuclei did not become stable until a few minutes had elapsed. When the universe was younger than this, it consisted of a very dense mixture of particles and antiparticles, and many different types of particles would have been present.

An analogy for the young, expanding universe is helpful. Suppose we were to heat an oven on a cold winter's night so that its temperature exceeded one hundred degrees. If we were then to place some steam into this oven, it would remain evaporated as long as we kept the heat turned on. Consider what would happen if we then disconnected this oven and took it outside to where the temperature was well below freezing. Clearly, the oven would begin to cool as it lost heat to the surroundings. As the oven temperature dropped, the steam would eventually condense back into water, and, finally, when the temperature had become sufficiently low, the water would freeze into ice.

The successive transitions from steam to water and from water to ice would occur quickly once the temperature became low enough. The time taken to complete each transition would be significantly *less* than the time taken for the oven to cool. The steam would remain evaporated for quite some time until the critical temperature of one hundred degrees had been reached. At this point, it would rapidly condense into water. The water would then remain as a liquid until the temperature had dropped to the freezing point. Only then would it transform into ice. The time taken to complete this second transition would be far less than the time required for the temperature of the oven to drop from one hundred degrees.

In short, the steam condenses into water, and this change is followed by a period of relative stability until the water freezes into ice. This means that the history of the water inside the cooling oven may be separated into three distinct eras corresponding to when the water was in a gaseous, a liquid, or a solid form.

The evolution of the early universe may also be discussed in terms of separate eras. These eras are defined by the specific properties exhibited by the matter at the time. The matter within the expanding universe cools in much the same way as the steam inside the oven. As the universe expands, its temperature drops and eventually reaches a critical value, causing the properties of the matter to change in some fundamental way. The change occurs very quickly and the overall effect is that the matter behaves differently afterwards, just as ice and water behave differently.

A number of fundamental events occurred during the big bang that separated these eras from one another. We will discuss these events in chronological order.

The first significant event in the history of the universe would have been its creation. In principle, this point may be employed to define the zero of time. At this stage in our discussion, however, we cannot say how the universe might have come into existence. The reason is that quantum fluctuations associated with the force of gravity would have been important. But we do not yet know how to incorporate these effects into the picture. The physical processes and theories that we have discussed in the preceding chapters break down when one attempts to include them. Our theories, at least in their present form, are too restrictive.

Consequently, we are limited in how far back we can go with the physical insight that we have acquired thus far in our discussion. The limit is the first 10^{-43} seconds of the universe's history. We cannot discuss what might have happened before the universe was this old. This time scale is known as the *Planck time* after the pioneering quantum physicist Max Planck.

In Chapter 12, we will explore how our theories can be extended to include quantum gravitational effects. This will lead us to an improved understanding of the origin of the universe. For the purposes of the present discussion, though, we will trace the history of the universe from the first 10^{-43} seconds.

The superstring era began when the universe was this old. The distance light can travel during the Planck time is only 10^{-35} metres. This scale is known as the *Planck length*. Since nothing can travel faster than light, the Planck length represents the size of the observable universe at this time. In view of this, the superstring era is sometimes referred to as the Planck era.

At the start of the superstring era the temperature of the universe was 10^{32} degrees. This is the critical temperature at which the four forces of nature are thought to become unified into the superforce. The stringy nature of matter also becomes apparent at these energies. There was only one force in operation at the Planck time, and this acted between the superstrings. The superstring theory predicts that the universe should have at least nine dimensions of space. These dimensions would have had comparable sizes at this stage.

The superstring era drew to a close when the superforce split into the force of gravity and the Grand Unified (GUT) force. This event defined the beginning of the Grand Unified era. Gravity began to act as a separate force. Only three of the spatial dimensions expanded, however. The reason for this is not fully understood, although we will discuss one possible explanation in Chapter 10.

The tension in the superstrings increased as the temperature fell. The falling temperature caused the strings to shrink, and they began to resemble the pointlike objects that we identify today as elementary particles and antiparticles. They continually collided into one another, because there was very little room for them to move around. The universe at this time may be viewed as a hot, dense soup of particles and antiparticles.

Even though the temperature of this 'soup' was falling, it was still sufficiently high that the quarks and the leptons were able to exchange *X* particles. These particles were responsible for mediating the GUT force. This caused the quarks to decay into leptons, and vice versa. Quarks and leptons were effectively indistinguishable at this stage in the universe's history.

As the expansion of the universe continued, the temperature kept falling. Once the temperature fell below a critical value, the exchange of *X* particles between the quarks and leptons became highly improbable, although it could still proceed in an exceptionally violent collision.

The quarks and leptons were then no longer able to significantly influence each other by means of the GUT force. Subsequently, each particle found it extremely difficult to decay into the other. The GUT force effectively ceased to operate, and this defined the end of the Grand Unified era. This critical stage was reached when the universe was about 10^{-35} seconds old. The temperature at this time was 10^{27} degrees. The GUT force split into the strong and electroweak forces, and this heralded the beginning of the electroweak era.

The electroweak era lasted for about 10^{-10} seconds. At the beginning of this era the temperature was still so high that the gluons were unable to bind the quarks together. The quarks behaved as free particles during this era. The *W* and *Z* particles initially behaved in a way similar to the massless photons, because their kinetic energy dominated any energy associated with their mass. These particles were essentially indistinguishable from the photons, so the electromagnetic and weak forces remained unified.

As the temperature fell, the average collisions between the particles were less energetic. The masses of the *W* and *Z* particles became important once the temperature had dropped sufficiently. The key physical condition for this to occur was that the temperature of the universe fell below a critical value. The weak force and the electromagnetic force then began to affect the particles in different ways. When the universe was 10^{-10} seconds old a transition occurred as the electroweak force split into the weak force and the force of electromagnetism. The corresponding temperature at that time was 10^{15} degrees. The electroweak era came to an end at this point. From then on, the development of the universe was governed by the four forces of nature that exist today.

The electroweak era was followed by the quark era. This era lasted until the universe was about 10^{-4} seconds old. Quarks interact with other quarks via the strong force. As we saw in Chapter 5, the defining feature of the strong force is that it becomes *weaker* on smaller scales. At the start of this era the temperature of the universe was so high that the quarks still behaved as if they were free particles. They gradually lost energy as the universe cooled, and the strong force became more influential as the temperature fell. Eventually, the gluons were able to hold the quarks together, and the quarks became confined into groups of two and three. This occurred when the temperature had fallen to around 10^{12} degrees. After this time, quarks could not exist in the universe as single particles, and the quark era came to an end.

The quarks effectively froze together. A similar process occurs when steam condenses into water. An individual water molecule may move around freely when the temperature is above one hundred degrees. Below this temperature, it does not have sufficient kinetic energy to resist the attractive forces that operate between it and the other molecules. Hence, the molecules bind together in the form of a liquid. Likewise, the quarks became confined in the early universe, although they only joined with one or two other quarks at a time.

The quarks that grouped into pairs soon decayed, as did the heavier quarks contained within the triplets. The triplets that were composed of up quarks and down quarks were stable. Those that contained two up quarks and one down quark formed protons, whilst those containing two down quarks and one up quark formed neutrons. Thus, the neutrons and protons that exist today in the nuclei of atoms were formed shortly after the quarks became confined, that is, when the universe was about 10^{-4} seconds old.

The temperature of the universe when these particles formed was still very high. Indeed, it was much higher than the current temperature at the surface of the sun. The strong force was unable to bind the protons and neutrons together to make nuclei, and they behaved initially as free particles. If a nucleus somehow did manage to form, the photons that were also present readily smashed it apart when they collided with it.

The universe cooled as it expanded, and the production of atomic nuclei eventually became possible. This production of nuclei is known as *nucleosynthesis*. This entire process was completed after about three minutes. The temperature of the universe during nucleosynthesis was about one billion degrees.

The simplest nucleus is that of hydrogen; this is just a proton particle. In a sense, this nucleus was formed when the quark era came to an end. When a neutron binds with a proton, the result is a deuterium nucleus. In principle, more complicated nuclei may then be produced. It was at this time that helium nuclei, which consist of two protons and two neutrons, were formed. More precisely, this helium nucleus is known as helium-4, because it contains a total of four particles. There is also a helium nucleus known as helium-3. This nucleus consists of two protons and a single neutron particle.

A key feature of the helium-4 nucleus is that the heavier nuclei that are formed by attaching a further proton or neutron to it are extremely unstable. This implies that it is very difficult to form such nuclei in appreciable numbers. For this reason, it is also very difficult to form much heavier nuclei by further synthesis. Yet attaching a deuterium or helium-3 nucleus to helium-4 does liberate a certain amount of energy in the form of gamma rays, and these reactions are therefore energetically favourable. They result in the production of a small fraction of lithium and beryllium. Lithium and beryllium nuclei contain three and four protons, respectively.

The universe continued to cool during the synthesis of these light nuclei. By the time this synthesis was over, there was insufficient energy available to produce the much heavier nuclei. For example, the protons that had remained free did not have enough energy to become attached to the helium nuclei, and the photons lacked the power to break these nuclei apart. Consequently, the significant production of nuclei heavier than helium-4 did not occur during the big bang, although as we just mentioned, some lithium and beryllium was produced. This is important because the heavier nuclei that are essential to life – such as carbon and oxygen – were *not* formed at this stage.

As we shall see in the following chapter, more protons than neutrons were present in the universe when nucleosynthesis took place. Most of the neutrons went into helium-4 nuclei, and much smaller fractions went into deuterium, helium-3, lithium and beryllium. The excess protons remained as hydrogen nuclei.

The universe continued to expand after nucleosynthesis was completed, but nothing significant happened for a further three hundred thousand years or so. By this time, the temperature had dropped to three thousand degrees. This was sufficiently low for the electrons and nuclei to form neutral atoms. The electrons had been present in the universe up to this point, but they had not played a significant role due to their relatively low mass. The formation of atoms defined the beginning of the matter era.

What happened during the matter era is similar to what transpired during the nucleosynthesis era. Recall that when the temperature was above one billion degrees, the electromagnetic radiation had enough energy to split apart any nuclei that may have formed. Likewise, if an atom somehow managed to form when the temperature was above three thousand degrees, the radiation would have soon collided with it and knocked the electrons free. Below this temperature, however, the radiation lacked sufficient energy to free the electrons, and the atoms could survive.

The formation of atoms was a very significant event in the history of the universe. Electromagnetic radiation interacts only with electrically charged objects; it does not influence electrically neutral ones. Before the atoms had formed, the negatively charged electrons and positively charged nuclei had interacted with the radiation.

Atoms are electrically neutral because they contain an equal number of protons and electrons. The formation of atoms resulted in the disappearance of naked electric charge in the universe. The radiation then found it extremely difficult to interact with the matter. Only radiation with just the right energy to cause electrons to jump between energy levels was affected. This was a very small fraction of the total amount of radiation present. The matter and radiation effectively *decoupled* from one another. (We should remark that the decoupling of matter and radiation was not exactly instantaneous, but the precise details will not affect our subsequent discussion.)

Since most of the radiation no longer had any free charged particles to interact with, it remained essentially undisturbed as the universe continued to expand. Indeed, it has been able to survive intact through to the present era. In the next chapter we will see that the existence of this radiation in the universe today provides strong support for the big bang model.

The fundamental nature of the matter in the universe did not change significantly once the atoms had formed. The appearance of atoms therefore represents the final transition in the history of the early universe. In a sense, this may be viewed as the point where the big bang ended.

After the big bang was over, the universe continued to expand, but the expansion was not precisely uniform. This is to be expected. In the extremely dense environment of the early universe, the matter would have been distributed somewhat randomly, and some regions of the universe would have been slightly denser than others.

A higher density in a given volume of space implies that more matter is present than in the surrounding regions. Consequently, even more matter would have been attracted from the surroundings into these high-density regions. The regions of high density became even denser, whilst the surroundings became effectively devoid of matter. Eventually, the gravitational attraction of the matter in the denser regions became more important than the outward effects of the cosmic expansion.

In effect, these denser regions formed into separate 'islands' of matter that then collapsed because of their own self-gravity. There would have been local regions in each island where the density was slightly higher than the average. These regions would have attracted more matter in the manner just described. They would have collapsed at a faster rate than the surrounding regions. Each island therefore split into separate mini-islands as matter flowed into the areas of higher density.

The matter in each of these mini-islands would have heated up considerably as it became compressed into a progressively smaller volume. Indeed, the temperature at the centres of the mini-islands eventually became sufficiently high for hydrogen nuclei to fuse into helium nuclei. As we discussed in Chapter 2, this process of 'hydrogen burning' occurs in the centre of stars. It released a considerable amount of energy in the form of gamma rays. These gamma rays provided an outward pressure that acted on the outer regions of the mini-islands. A feedback mechanism was established whereby the inward pull of gravity was precisely balanced by the outward pressure of the gamma rays.

These mini-islands formed into stars, and the original islands of matter represent the galaxies that we observe today. This process of fragmentation took somewhere in the region of one billion years to complete.

Our sun, however, did not form out of one of these mini-islands of matter. We saw that elements heavier than helium, such as carbon and oxygen, could not have been produced during the big bang because the helium nucleus is so stable. Thus the first stars to form in the universe could not have contained these elements. But the sun does contain traces of the heavier elements. How, then, did stars such as the sun form?

We may answer this question by considering what happens to a typical star over time. A star is stable as long as enough radiation is produced in its core to maintain the necessary outward pressure. We know that stars shine and that this means that some of the radiation produced in the nuclear reactions must escape. If the sensitive balance inside the star is to be maintained, this radiation must be continually replaced. To replace this radiation requires the conversion of more hydrogen into helium. The amount of available hydrogen within the star is limited, and the core gradually becomes dominated by helium. The production of fresh radiation drops drastically when most of the hydrogen has been converted. The outward pressure from the radiation then decreases, and this allows a new phase of collapse to begin.

The mass of the star determines what happens during the subsequent collapse. The matter inside the star consists mainly of protons, neutrons and electrons. As we saw in Chapter 5, both the proton and neutron consist of three quarks, and these particles are examples of fermions. The electron is also a fermion. The defining feature of these particles is that there exists a limit to how many of them can be confined into a given region of space.

This has important consequences for what happens to the star once it has started to collapse. We may view the electrons in the collapsing star as glass marbles contained within a bag. In this analogy, the collapse of the star corresponds to a shrinking of the bag. The marbles begin to move closer as the volume of the bag decreases. Eventually, the marbles will be unable to move about freely because no room will be left in the bag. Since these marbles are hard objects, they cannot easily be deformed. They will become wedged together into a lattice once the volume of the bag has become sufficiently small. This lattice will produce a new outward pressure that is very strong, and it will become extremely difficult to shrink the bag further.

The electrons inside the collapsing star behave in a way similar to the marbles inside the bag. A point is reached in the collapse when there is simply not enough room for the electrons to move around.

This results in a new pressure being established, and in some stars, this pressure is able to halt further collapse.

In 1928 the Indian astrophysicist Subrahmanyan Chandrasekhar considered how electrons might affect the collapse of burnt-out stars. Similar calculations were also performed independently by the Russian physicist Lev Landau. Both Chandrasekhar and Landau realised that the electrons would halt the collapse of a star that was not too massive. Such a star ultimately collapses into what is known as a *white dwarf*. A white dwarf has a size comparable to that of the Earth, but its density is very much higher.

What would happen to more massive stars? The inward force of gravity increases as the mass of the star increases. Is it possible that gravity might be strong enough in these larger stars to overcome the pressure of the electrons?

Let us return to our picture of the marbles inside a bag. Suppose we had access to a very strong compressor. In principle, we could apply this compressor to the bag in an attempt to squash it further. If the force of the compressor was sufficiently strong, the glass would shatter, and the lattice structure maintained by the marbles would be broken. A similar process occurs in the collapse of more massive stars. The force of gravity overcomes the pressure generated by the electrons, and these particles are unable to halt the collapse.

There is a critical mass above which gravity is able to dominate the electron pressure. It is about one and a half times the mass of the sun. A star such as the sun will eventually form a white dwarf when it runs out of nuclear fuel in about five billion years time. More massive stars continue to collapse and do not form white dwarfs.

What happens to these massive stars? Protons are also fermionic particles, so they too behave rather like marbles inside a bag. We might expect these particles to provide a new pressure that is similar to that produced by the electrons, but this does not happen. The protons and electrons become tightly packed inside the collapsing star, and their average separation soon becomes less than the distance over which the weak force can operate. This force is able to combine protons and electrons and turn them into neutrons. Such a transformation occurs within the collapsing star. Indeed, so many neutrons are produced in the core that it becomes saturated with them.

This conversion has important consequences. Since the neutrons are also fermionic, their marblelike properties produce a new pressure that prevents further collapse. The outer regions of the star are then ejected, and a huge amount of energy is released in the process. This causes the brightness of the star to increase by many orders of magnitude. When viewed from a distance, the star appears to explode, and this event is referred to as a *supernova*. It is during this time that the very heavy elements are produced because of the extreme temperatures that are attained.

Many of the original stars would have been sufficiently massive that they exploded as supernovae within one billion years of forming. The ejected remains of these stars cooled over time. Some of the matter would have eventually collapsed around a region of high density. One such region collapsed into what is now our solar system, with its sun and planets, about five billion years ago. In this scenario, therefore, the carbon and oxygen in our bodies was synthesized in the core of a massive star and then released in a supernova explosion. In a very real sense, each of us is made from stellar material!

This completes our account of the history of the universe. We have followed its progress from the first 10^{-43} seconds through the billions of years during which galaxies, stars and our own solar system were formed. However, this is not the full story by any means. We have yet to discuss the cosmological observations that indicate that the universe did indeed develop out of a big bang. Although there is strong support for the picture we have discussed in this chapter, it may be incomplete. In the next chapter, we will discuss why the big bang model may need to be extended.

8

Beyond the Big Bang

In Chapter 2, we discussed the very reliable observations that show that the universe is expanding at the present era. The current expansion of the universe represents the first argument in favour of the big bang model, although it is not proof by itself, as explained in Figure 4.3. Two other important observations support the model. The first is the relative abundances of hydrogen and helium in the universe. The second is the existence of cosmic radiation at the present era.

The amount of helium that was produced during the nucleosynthesis era of the big bang was determined by the relative numbers of neutrons and protons that were present at that time. These particles were formed shortly after the quark era, when the quarks became confined by the strong force. The universe was about 10^{-4} seconds old when this occurred. It took another three minutes or so for the universe to cool sufficiently for the synthesis of helium to be completed.

Thus, the neutrons and protons had to wait before they could begin to produce atomic nuclei. Because the mass of the neutron is slightly higher than that of the proton, the neutron has slightly more energy, and a free neutron may decay into a proton. What happens is that a down quark changes into an up quark, and this transformation is made possible by the weak interaction. (An electron is also produced to conserve electric charge, and a lepton known as an 'antielectron neutrino' also appears in the decay, but these details will not concern us here.)

The temperature of the universe at the time the neutrons and protons first formed was still relatively high. The reverse reaction, whereby a proton changed into a neutron, was also possible via the weak interaction. A state of equilibrium between the neutrons and protons would have been established, and equal numbers of each would have been initially present in the universe. As the temperature dropped, however, the neutron's higher mass became more important, and protons were no longer able to transform into neutrons. This occurred when the universe was about one second old, and the temperature was 10^{10} degrees. The neutrons were still able to decay into protons even though the temperature remained too high for helium to form. As more and more neutrons decayed, the fraction of protons in the universe increased.

The overall result was that more protons than neutrons were present in the universe when nucleosynthesis eventually took place. The majority of surviving neutrons became locked into helium-4 nuclei, whereas the remainder ended up in deuterium, helium-3 and lithium. Many free protons were left, and these ultimately went into forming hydrogen atoms.

The nuclei of both hydrogen and helium are very stable. Thus, most of the nuclei that formed during the big bang have survived to the present day. It is this feature that allows us to test the validity of the big bang picture. The idea is to employ our understanding of nuclear physics to determine how much hydrogen and helium should have formed during the big bang. These relative amounts should then have remained constant throughout the subsequent evolution of the universe. We are therefore able to *predict* how much hydrogen and helium should be present in the universe today. We may then compare this prediction with the amount that is actually observed. A good agreement between the theoretical prediction and the observations would provide strong support for the big bang model.

When the calculations are performed, the theory predicts that the fraction in mass of helium should be about twenty-five percent. How does this compare to the observations? The observations are complicated somewhat by the fact that helium is also produced in stars when hydrogen nuclei fuse together. If the star subsequently explodes in a supernova, this helium will be ejected into outer space. The stellar helium would be indistinguishable from the helium produced in the big bang and could contaminate the observations. The observations must be made in regions where there are no stars.

Remarkably, the observations in these regions agree very closely with the prediction made from the big bang model. We can observe this agreement in many different places throughout the universe. If the helium had been produced due to local processes, such as in stars for example, we would expect the concentration of helium to vary from place to place. That this is *not* observed suggests that the helium indeed has a primordial origin. This agreement between theory and observation implies that our picture of the early universe when it was a few minutes old is reasonably accurate.

The universe continued to expand and cool after the nucleosynthesis era. It remained in a relatively stable state for another three hundred thousand years or so until atoms were able to form. During this time, the universe consisted of a hot plasma of positively charged nuclei, negatively charged electrons and photons. Because photons interact with charged particles, they collided with the electrons and nuclei. Many such collisions took place each second because the particle densities were still very high.

The photons decoupled from the matter when atoms formed. When this separation was over, there were no longer any free, charged particles present in the universe. The photons were now able to travel through the universe largely unhindered. They have survived to the present day.

How did the expansion of the universe affect the energy of these photons? As we discussed in Chapter 4, the cosmic expansion arises because space itself is being stretched. This process is similar to the stretching of a balloon when it is pumped with air. It is important to emphasize that these photons filled the entire universe at the start of the matter era. We can incorporate these photons into the analogy by drawing over the entire surface of the balloon with a marker pen. As the balloon is blown up, its surface will remain completely covered by the markings of the pen. This implies that the photons will still fill the universe after decoupling. However, since the same amount of ink is being employed to cover a greater surface area, the intensity of the ink will fade. Likewise, the energy of the photons will fall as the universe expands. The decrease in the intensity of the ink is the same over the entire balloon. This implies that although the photons lose energy, they all do so at the same rate.

The big bang model predicts that the universe should be bathed in photons – that is, electromagnetic radiation – at the present era. These photons should all have the same energy. Today this energy is quite low, since the universe has been expanding for at least ten billion years. Indeed, modern cosmological theory predicts that the current temperature of the radiation should be only three degrees above absolute zero.

The wavelength of this radiation is relatively long, much longer than the wavelengths associated with visible light, for example. The big bang model predicts that the wavelength should presently be in the range of a few millimetres to a few centimetres, which corresponds to the *microwave* region of the electromagnetic spectrum. The entire universe should currently be bathed in microwaves if the big bang model is correct. This radiation is referred to as the *cosmic microwave background*. It is of the same type that is employed in microwave ovens. However, we are not in danger of being cooked by this radiation, because its intensity is extremely low.

The cosmic microwaves were discovered in 1965 by two researchers, Arno Penzias and Robert Wilson. The existence of this radiation coming from every direction in the universe provides very strong support for the view held by most cosmologists that the big bang model is essentially correct. However, there are still some features of the model that are not fully understood. We will conclude this chapter by investigating these issues further.

Let us begin by reconsidering the expansion of the universe. In Chapter 3 we established that the universe is currently expanding. The question that arises is whether this expansion will continue. That is, will the universe expand indefinitely, or will it begin to recollapse at some point in the future?

The expansion of our universe can be modelled by considering what happens when a ball is thrown into the air. In this analogy, the height of the ball above ground level represents the size of the universe. In this sense, a greater height corresponds to a larger universe. A ball that is rising then coincides with an expanding universe. A contracting universe is represented by a ball that is falling.

Consider what happens to the ball after it has been thrown by an athlete, say, who gives the ball a certain amount of kinetic energy. This enables it to move upwards against the pull of gravity. The attractive nature of the gravitational force may be viewed as negative energy. As the ball rises, it loses kinetic energy and gains potential energy.

Even if the throw is particularly good, the ball will rise only a few metres before falling swiftly back to Earth. This follows since the ball is given only a small amount of positive kinetic energy by the thrower.

The ball will reach a greater height if it is given more kinetic energy. This could be achieved, for example, by firing it from a cannon. If the cannon is sufficiently powerful, the ball will completely escape from the gravitational pull of the Earth, because its initial energy is so high. In this case, the ball will never fall back to the surface of the Earth, and its height will effectively increase forever.

The former example, where the ball is thrown by the athlete, corresponds to a universe that expands initially but then recollapses after a certain time. The second case, where the ball is shot from a cannon and escapes, corresponds to a universe that is able to expand forever.

The key point is that there are two factors that influence the maximum height attained by the ball. The first is the initial kinetic energy given to the ball when it is released. The second is the effect of gravity at the Earth's surface. The influence of gravity is determined by the mass of the Earth. The future behaviour of the universe is governed in a similar way. The initial conditions at the big bang cause the space between the particles to stretch. The expansion slows down due to gravity, and the rate of decrease is determined by the amount of matter that is in the universe.

These two effects are continually battling each other. The answer to whether the universe will expand or contract in the future is determined by which effect will ultimately dominate the other. If enough expansion energy was available at the big bang, the universe will be free to expand indefinitely. If there is too much matter in the universe, the negative energy associated with gravity will eventually cause the expansion to give way to a contraction.

The different possibilities are shown in Figure 8.1. In this picture a line labelled as D is shown. This line applies when the initial energy of expansion is *precisely* balanced by the negative gravitational energy in the universe. It corresponds to the case where the ball has just enough kinetic energy to escape from the surface of the Earth. In this case, the universe can expand forever.

Which of the lines in Figure 8.1 represents our universe? Rather surprisingly, observations indicate that our universe lies very close to the line D. Indeed, it is currently so close to it that we cannot say whether it lies above or below the line. The uncertainties in the observations are still too large for us to reach a definite conclusion. We do not yet know whether the universe will expand forever or will ultimately recollapse. This has significant implications for our understanding of the big bang.

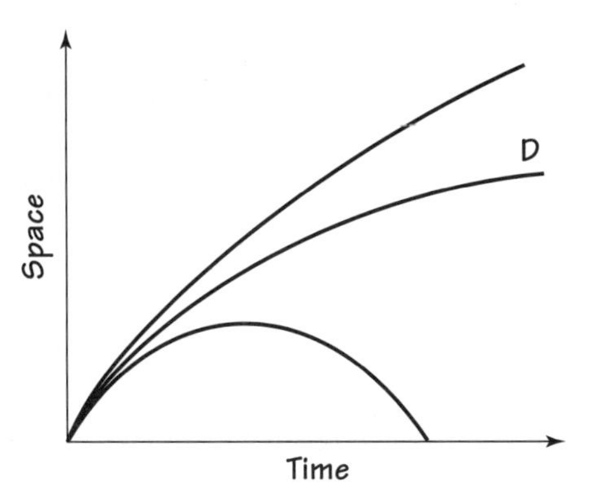

Figure 8.1. The future development of the universe is determined by the relative amount of matter that is present. If there is enough mass in the universe, the expansion will eventually cease and the universe will recollapse. Alternatively, the universe may expand indefinitely if the density of matter is sufficiently low. The two possibilities are separated by the line *D*. This corresponds to the case where the universe is just able to avoid recollapse.

Figure 8.1 has an important feature. Notice how the three lines never intersect. The lines move further and further away from each other as time proceeds and converge as we go back in time. This implies that the line representing the expansion of our universe will gradually move away from the line *D* as the universe ages, but it would have been closer to it during the big bang.

The problem is that the universe is still very close to *D* today even though it has been expanding for over ten billion years. This implies that it must have been *extremely* close to line *D* at the big bang. In other words, the universe must have started out with its positive energy almost precisely balanced by its negative gravitational energy. The two must have been almost identical, otherwise the universe could not have remained so close to line *D* for such a long time.

We have seen that there is a limit to how far back in time we may go with our current theories. The earliest time we can consider is the Planck time, corresponding to 10^{-43} seconds after the creation of the universe. It is possible to calculate how close the balance between the different energies must have been at this time. The relative difference between the two could not have been more than 10^{-60} if consistency with the observations is to be maintained. In other words, the magnitude of the one exceeded the other by no more than one part in 10^{60}.

This is a remarkably *small* number. The big bang model in its present form is unable to explain how such a precise balance arose. This shows that the model is incomplete in some way. There must have been other physical processes in operation at early times that pushed the universe towards the line *D*. We will examine these processes in the next chapter.

A second puzzle associated with the big bang model is related to the process of galaxy formation. We know from our observations that billions of galaxies exist in the universe today. The question of how these galaxies formed out of the primordial fireball is one of the great unresolved issues in cosmology. A considerable amount of time and effort is currently being devoted to answering this question.

In the previous chapter we outlined a scenario that describes, at least in principle, how galaxies may have begun to form once atoms became stable. The fundamental idea is that small irregularities in the distribution of matter were produced during the big bang. The density of these regions gradually increased after decoupling as matter was attracted towards them. Eventually, the attractive force of gravity within these regions became so strong that they began to behave as gravitationally bound objects. These islands of matter then began to collapse. In the process, they fragmented into numerous mini-islands, and these mini-islands eventually formed into stars.

Unfortunately, there is a problem with this picture of galaxy formation. The problem lies in the initial size of the density irregularities. These irregularities determine how quickly matter is attracted onto the island regions. A larger irregularity attracts matter at a faster rate, since the gravitational pull is stronger. Conversely, less mass is attracted towards a smaller irregularity.

Our current understanding of the big bang allows us to calculate the expected size of the initial irregularities. The young universe behaved like a hot plasma, and the processes that occur inside such a plasma are relatively well understood. A reliable estimate for the magnitude of the irregularities can therefore be derived. The results are rather surprising. The fluctuations are predicted to be very small. This has profound consequences for the picture we discussed earlier in that it implies that the formation of galaxies would have been a very slow process. Indeed, there would not have been enough time for all of the structure that we observe today to develop. This suggests that the initial perturbations must have been generated by a mechanism that we have yet to discuss.

The final problem with the big bang model that we will discuss in this chapter is related to the cosmic microwave background radiation. This radiation has remained essentially undisturbed since the universe was about three hundred thousand years old. It exhibits a surprising feature. It has the same temperature in all directions to one part in one hundred thousand. This finding implies that this background radiation currently has the same temperature throughout the entire observable universe. What is the significance of this result?

Any distribution of matter or radiation with a uniform temperature is said to be in a state of *thermal equilibrium*. The fact that the temperature of the cosmic microwave radiation is uniform tells us that it must have been in thermal equilibrium when it decoupled from the matter. It takes a finite amount of time for equilibrium to be established. Surprisingly, the big bang did not last long enough for all of the radiation that we observe today to acquire the same temperature. A problem then arises when we attempt to explain why the temperature does not vary from place to place.

How is thermal equilibrium established? Suppose we take a cup of hot water and pour it into a cup of cold water. The temperature of the water is a measure of how fast the molecules are bouncing around in the cup. Initially, the temperature will be unevenly distributed throughout the cup, so the water will not be in thermal equilibrium. Over time the molecules from the hot cup will bounce into those from the cold cup, which will cause some of the excess energy to be transferred to the cooler regions. This redistribution of energy will continue until the temperature is the same throughout the cup. From then on the water will be in thermal equilibrium.

The important point is that it takes *time* for the heat to become evenly distributed throughout the cup. The water molecules at the top and bottom do not immediately reach the same temperature. Likewise, it would have taken time for the radiation that we observe today to reach thermal equilibrium.

We may understand why there was insufficient time to establish thermal equilibrium during the big bang by considering a one-dimensional analogy. Let us view the universe as a piece of elastic, as shown in Figure 8.2. Our current position in the universe is denoted by the point O. At any given time during the universe's history, there is a limit to how far an observer can see. This follows because the universe has a finite age. Consequently, a light signal that began its journey shortly after the big bang could only have travelled a finite distance. Since nothing can travel faster than light, this is the maximum distance accessible to observations.

The farthest observable regions in our universe are denoted in Figure 8.2 by the lines *A* and *B*. These lines are located in opposite

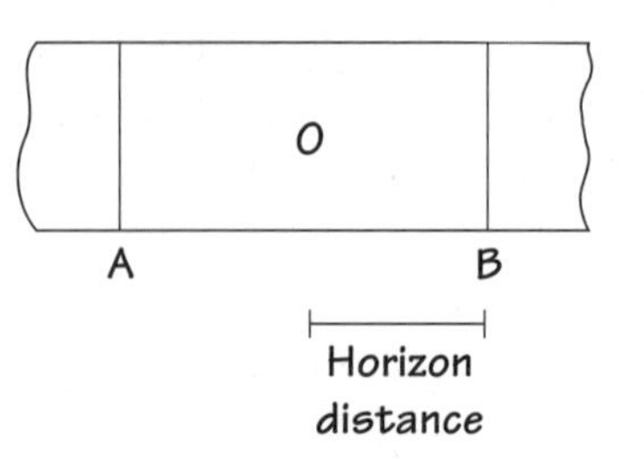

Figure 8.2. The finite age of the universe implies that there is a limit to how far an observer at *O* can see. This limit is denoted by the lines *A* and *B*. The distance to these lines is referred to as the horizon distance and represents the size of the observable universe. The actual universe may extend beyond the horizon distance.

directions to our line of sight. The distance between us and these lines is referred to as the *horizon distance*, and it is determined by how far light has travelled since the beginning of the universe. For any given time the horizon distance is arrived at by multiplying the age of the universe by the speed of light. It is currently about ten billion light years. This distance also represents the maximum scale over which physical processes can operate.

Although the horizon distance determines how far we can see at a specific era, it does not necessarily represent the actual size of the universe. The elastic may continue beyond the horizon distance, or it may not. We have no way of deducing the real extent of the elastic from our observations, because there has not been enough time since the beginning of the universe for a photon sent from beyond the horizon to reach us.

Let us consider the era in which the matter and radiation decoupled from one another when the universe was three hundred thousand years old. The physical distance separating lines *A* and *B* would have been shorter at decoupling. In general, physical distance in the universe may be thought of in terms of the distance that separates two typical particles, as we discussed in Chapter 1. The corresponding horizon distance at the time of decoupling would also have been smaller than it is today, because light had had less time to travel. The question that arises is whether the presently observable region of the universe that is bounded by the lines *A* and *B* in Figure 8.2 was larger or smaller than the actual horizon distance at that time.

The two possibilities are shown in Figure 8.3. In Figure 8.3a the distance between *A* and *B* exceeds the horizon distance, but the reverse is true in Figure 8.3b. In the former case, points *A* and *B* would have been unable to communicate with each other, and their temperatures should

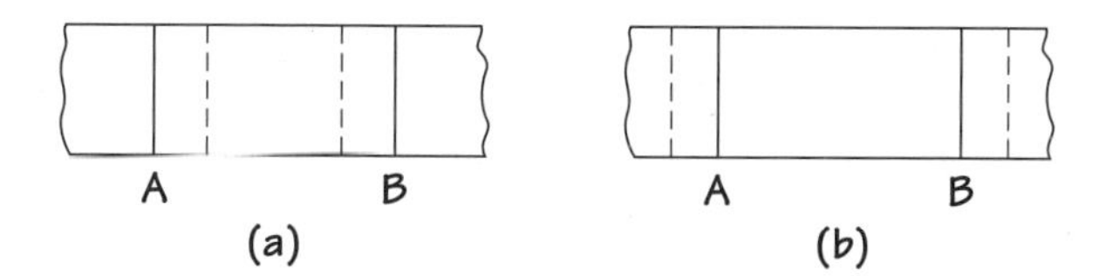

Figure 8.3. (*a*) A schematic diagram of the universe at the era of decoupling. The dashed lines denote the size of the horizon at that time. The region of the elastic (universe) bounded by the lines *A* and *B* corresponds to that part of the universe that is contained within our current horizon. In this diagram, the horizon size at decoupling is smaller than the distance between *A* and *B*. These two regions would not have been able to communicate with each other, and there is no reason to suppose that they should have the same temperature. (*b*) If the horizon size is greater than the distance between the two regions, this is not a problem.

have remained uncorrelated. But they would have had enough time to acquire the same temperature via physical interactions if Figure 8.3b applied.

Which of these two possibilities corresponds to our universe? The rate at which the physical distance between *A* and *B* increases after decoupling depends on the amount of expansion energy that is available in the universe. In the big bang model, physical distances grow at rates that are *less* than the speed of light, in the sense that the space between two average particles stretches at a rate that is slower than light-speed. The horizon distance, however, grows *at* the speed of light.

The physical distance between lines *A* and *B* in Figure 8.2 corresponds to the horizon distance at the present time. Thus, the horizon distance at decoupling must have been *smaller* than the physical distance between *A* and *B*, since the latter grows at a slower rate. Figure 8.3a therefore applies and the photons at *A* and *B* could *not* have been in direct contact with each other at decoupling.

A problem then arises when we attempt to understand why they both have the same temperature. How had these photons managed to reach a state of thermal equilibrium by the time of decoupling? It is as if two cooks had chosen to set their respective ovens to precisely the same temperature even though neither had any way of knowing what the other was doing.

To summarize, we have seen in this chapter how the big bang model can explain the observed expansion of the universe, the relative abundances of hydrogen and helium and the origin of the cosmic radiation. Despite these outstanding successes, a number of puzzling questions remain. Why is the expansion of the universe almost precisely balanced by the gravitational attraction of the galaxies? How did these galaxies form? Why is the temperature of the cosmic radiation uniform throughout the universe?

The big bang model is unable to answer these questions. It needs to be extended in some way. These problems might be resolved by physical processes that operated before and during the Grand Unified era when the universe was less than 10^{-35} seconds old. We will investigate the nature of these processes in the next chapter.

9

The Inflating Universe

In this chapter we will consider in what way the big bang model needs to be modified. Recall from the previous chapter that there are a number of shortcomings with the model, in particular, with the uniformity of the cosmic radiation. Let us reconsider this cosmic radiation further. Its uniform temperature is problematic because the model predicts that the horizon distance grows *faster* than the separation between two points in space.

Our assumption that the horizon distance grows more rapidly than the expansion of space may not always be correct in the environment of the very early universe. It is quite possible that space itself might have expanded faster than the horizon for a brief interval some time before the decoupling of radiation and matter took place. Suppose, for the moment that this had indeed been the case. How would it affect our conclusions regarding the cosmic radiation?

We could begin with a region of the universe that was much smaller than the horizon. Physical processes could have operated within this region to establish thermal equilibrium, and any temperature differences that might have existed could in principle be eliminated. If this region then grew at a faster rate than did the horizon, it would eventually come to exceed it.

The final result would correspond to Figure 8.3a when we identify the boundary of the initial region with the lines A and B. In this case, there is nothing peculiar about *A* and *B* having the same temperature, since they were originally so close to one another that energy could have been distributed evenly between them. Photons located at these points may have the same temperature without any problems arising.

In short, the observed uniformity of the cosmic radiation could be explained if space expanded faster than the horizon for some finite length of time during the big bang. This idea appears to have a problem, however. We know that the horizon grows at the speed of light. If space is to grow faster than the horizon, it needs to expand at speeds *greater* than that of light. Nothing is supposed to be able to travel faster than light, and such a rapid expansion seems to violate this principle.

How is this paradox resolved? The key point is that Einstein's theory actually says that no form of matter or radiation can travel faster than light *through space*. His theory says nothing about the speed with which space itself expands. The speed of the expansion of space is limited only by the amount of energy that is available to drive the expansion. There is no inconsistency as long as the matter that is travelling through this expanding space moves at speeds below the speed of light.

A rapid expansion of space would lead to a large growth in the volume of the universe in a very short time. We might say that the universe becomes 'inflated'. In view of this, such an expansion is referred to as *cosmological inflation*. We will shortly discuss how inflation might have occurred in the very early universe. Before doing so, however, let us consider whether inflation is also able to explain why our universe is currently so close to the line *D* of Figure 8.1.

During inflation the universe expands at a faster rate than in the big bang picture. Figure 8.1 must be modified when it is applied to an inflationary universe. All the lines in this diagram – including the line *D* – are shifted upwards. We have shown these modifications in Figure 9.1. Inflation begins at some time t_1 and ends at a time t_2.

The key point of Figure 9.1 is that the top and bottom lines move *towards* the line *D* during inflation. They will be closer to this line at any given time after inflation than they would have been if inflation had not occurred. In effect, the line representing our universe behaves *as if it was very close to D* during inflation, although it should be emphasized that it never actually touches *D*.

The longer inflation lasts, the closer the lines get to *D*. They then move away from it when inflation ends and the standard expansion

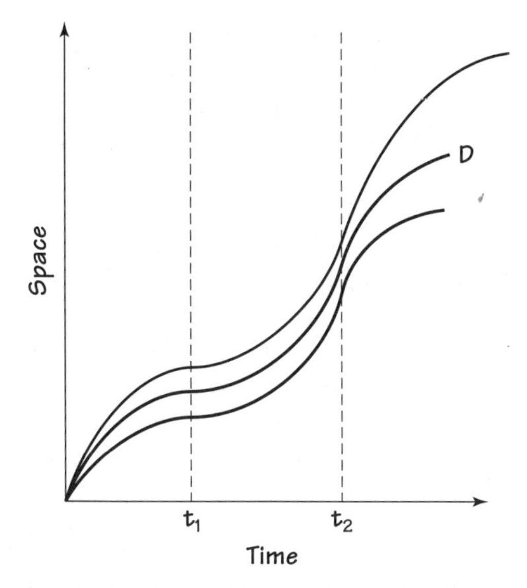

Figure 9.1. During inflation, the lines approach the line *D* very rapidly, regardless of whether they lie above or below it. The lines get nearer to *D* the longer inflation lasts. An inflating universe will behave as if its expansion energy is very nearly balanced by the gravitational attraction of its matter. The lines move away from *D* after inflation. If enough inflation has occurred, the lines may still be close to *D* a long time after inflation has ended. This is the case with our universe.

takes over. Thus, the line corresponding to our universe may still be very close to *D* at the present era if enough inflation occurred.

To summarize thus far, a period of inflationary expansion in the early universe can explain why the universe is so close to the line *D* at the present era and why the temperature of the cosmic radiation is so uniform. We have yet to address the question of where the tiny fluctuations necessary for galaxy formation may have come from. We will shortly see that these perturbations arose naturally during inflation because of quantum effects. Before we can investigate these effects, however, we must discuss what caused the young universe to inflate in the first place.

Let us return to our analogy from the previous chapter where we thought of the expansion of the universe in terms of a ball being thrown into the air. The height of the ball above ground level represented the size of the universe. Likewise, we may consider the *rate* of expansion of the universe in terms of how rapidly the ball continues to gain height once it is released.

The expansion rate of the universe at any given time after the big bang is determined by the initial conditions, in the same way that the subsequent behaviour of the ball as it continues its upward journey is uniquely determined by the amount of kinetic energy it acquires from the thrower when it is released into the air. As time proceeds, the ball loses kinetic energy and gradually slows down. In other words, it *decelerates*. Similar behaviour arises in the standard, hot big bang model. The expansion of the universe decelerates over time precisely because the energy density of the matter decreases too rapidly as the volume increases. This decrease is to be expected since the energy density is given by the energy of the matter divided by the volume.

It follows that a necessary condition for inflation to occur is that the energy density must not fall too swiftly. In view of this, let us suppose for the moment that the energy density of the universe remained almost *constant* at some very early time in its history. How might this alter the picture that we outlined in Chapter 7?

If the energy density is to remain constant, the energy that drives the expansion must increase in direct proportion to the volume of the universe. It would seem that we are already faced with a problem. The principle of energy conservation implies that energy cannot be created from nothing. Since the universe is isolated, we cannot acquire additional energy from outside. It must come from within the universe itself. What is the source of the extra energy that we require?

We saw in Chapter 5 how some forms of energy may be negative. The energy associated with the gravitational force is one example of negative energy. The key idea is that the increase in positive energy that results in the inflationary expansion is precisely balanced by the production of an equal amount of negative gravitational energy.

The total energy in the universe therefore remains constant, as required by the law of energy conservation. In a certain sense, the positive and negative energies originate from the vacuum. There is no specific creation of energy in this process, and the energy beforehand is the same as the energy afterwards.

We know that the universe is not inflating today, because the observed expansion of the galaxies is far too slow. The inflationary expansion must have eventually come to an end. This would have occurred when the energy density driving the inflationary expansion transformed into the energy associated with the masses and motion of ordinary particles.

When would inflation have occurred? We previously identified a number of eras associated with the big bang. These eras were defined by the type of matter that was present in the universe and by the different forces that were in operation. We saw how each era ended very quickly once the temperature of the expanding universe had fallen below a certain critical value.

It is possible that the inflation of the universe is associated with one of these transitions. This idea was first put forward in 1981 by Alan Guth, an American particle physicist. Guth had worked on Grand Unified theories and applied his knowledge of particle physics to the universe as a whole. He began by investigating what might have happened when the Grand Unified force split into the strong and electroweak forces.

Guth realized that the conditions in the universe at that time could have allowed inflation to occur in the manner that we described earlier. In Guth's original scenario, inflation began when the universe was 10^{-35} seconds old and ended when it was approximately 10^{-33} seconds old. The entire process took a fraction of a second to complete. During this time, the universe doubled in size every 10^{-34} seconds. It was a trillion times smaller than the size of a proton when inflation started, but had grown to about 30 centimetres by the time inflation had ended.

Unfortunately, Guth's original scenario did not work, because it resulted in a very inhomogeneous universe. This was inconsistent with observations of the cosmic microwave background radiation. The inhomogeneities arose because the inflationary expansion was associated with a phase transition in the early universe.

Two years later, in 1983, the Russian cosmologist and particle physicist Andrei Linde realised that inflation is a generic feature of many theories of particle physics. In particular, Linde showed that there was no need for a phase transition to occur during inflation, and, moreover, he argued that the temperature of the universe at the onset of inflation need not have been high.

This development represented a major deviation from the standard, hot big bang picture. Linde called his version of inflation the *chaotic* scenario. His fundamental postulate was that one should consider all possible initial conditions in the preinflationary universe and establish under what circumstances inflation arose. Since the expansion of the universe is so rapid during inflation, those regions that did inflate soon came to dominate the volume of the universe.

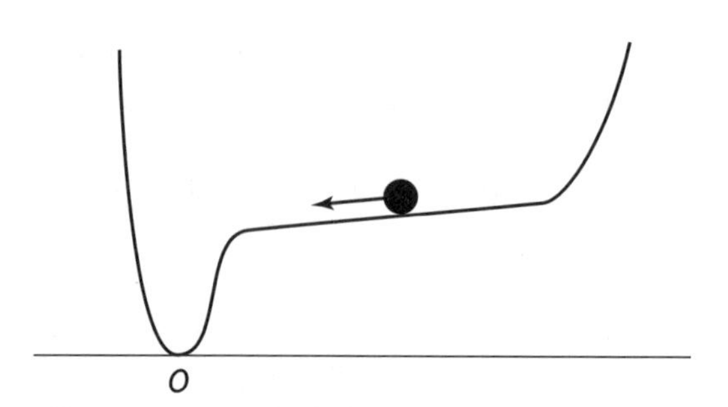

Figure 9.2. The energy density of the universe during inflation may be viewed in terms of the potential energy associated with a ball that rolls slowly down a hill. Inflation occurs when the ball is located on the plateau region and ends when it falls down the valley towards *O*.

The vast majority of realistic inflationary models that have been developed since Linde's breakthrough come under this general umbrella of chaotic inflation. We will discuss the physical processes that occurred during chaotic inflation in the remainder of this chapter and investigate some further consequences of the scenario in Chapter 10.

Consider the following analogy. The hill shown in Figure 9.2 has a number of important features. There is a valley at the bottom, as denoted by the point *O*. A plateau lies to the right of *O* and ends where it meets a steep cliff. This plateau is inclined slightly to the horizontal.

If we were to hold a ball at rest somewhere on the plateau, it would have a certain amount of potential energy associated with it. This would be determined by the ball's height above the point *O*. If we were then to release this ball, it would begin to roll slowly down the hill towards the valley. Its potential energy would be converted into kinetic energy in the process. Eventually, the ball would reach the edge of the plateau and then fall rapidly into the bottom of the valley. Its remaining potential energy would be quickly converted into kinetic energy.

The idea is that the potential energy of the ball represents the energy density in the universe. When the ball is located on the plateau, its potential energy remains almost constant, although it does fall slightly as the ball slowly rolls towards the valley. Thus, the energy density of the universe stays roughly constant as represented by this phase of the ball's journey, and this results in inflation.

Inflation ends when the ball reaches the edge of the plateau and rushes down towards *O*. The conversion of the ball's potential energy into kinetic energy represents the conversion of the energy density of the universe into particles. A vast amount of energy was released in this process, and the newly created particles were very energetic.

At this point the universe became very hot. Its temperature would have exceeded that associated with the electroweak phase transition and may even have been just slightly lower than the temperature of the GUT transition. Thus, the conditions in the universe immediately after inflation would have resembled those of the hot big bang. The history of the universe from this time onwards can then be described within the context of the big bang model.

The picture of inflation that we have just outlined is incomplete because it does not take into account quantum fluctuations. As we discussed in Chapter 5, intrinsic uncertainties arise in *all* physical processes, but these uncertainties only become important on very small scales. Since inflation occurred at a very early time, the universe would have been tiny in comparison to its present size. It is reasonable to suppose that quantum fluctuations would have played a significant role.

One consequence of these quantum fluctuations is that the energy and position of a particle can never be precisely measured. This same principle will apply to the ball as it travels along the plateau of Figure 9.2. We may understand how the inflationary process is modified by investigating how the motion of this ball is affected.

Suppose the size of this ball is comparable to that of a typical elementary particle. In this case, the uncertainties in the energy and speed of the ball are significant. These uncertainties will influence the motion of the ball by a tiny amount as the ball moves along the plateau. They operate in random directions, and sometimes the ball will be pushed slightly towards the cliff. On other occasions, it is directed downwards by the fluctuations. The position of the ball on the plateau will be uncertain to some extent. The overall result is that the ball may move towards the valley at a faster or slower rate than we previously thought.

What does this mean for the inflating universe? As we have seen, inflation proceeds when the ball is on the plateau, and it ends when the ball falls over the edge and rushes down into the valley. If the ball is slightly higher up the plateau, it will reach the valley at a *later* time than expected. Inflation will last a little bit longer. Conversely, inflation will end sooner if the quantum fluctuations push the ball downwards. In short, the overall effect of the quantum fluctuations is that they cause inflation to end *at different times in different regions of the universe.*

This time difference has important consequences. It means that different regions of the universe inflate by different amounts. Hence, the density of matter will vary throughout the universe after inflation. Some regions will be denser than others. The quantum fluctuations that act on the ball are very weak, so the density variations will be small. Nevertheless, they do play a significant role during the subsequent evolution of the universe. Indeed, they can lead to the formation of galaxies. The problem with the galaxy formation scenario that we discussed in Chapter 7 was that the expected fluctuations were too small in the standard big bang model. Surprisingly, the quantum fluctuations generated during inflation can produce irregularities of the required size.

This is a remarkable feature of inflation. It suggests that the largest structures in existence today may have arisen out of processes that occurred on the smallest of scales when the universe was just a fraction of a second old. Consequently, the idea of inflation can be tested experimentally. In principle, we may employ the theory to predict how the universe should look if galaxies grew out of these quantum fluctuations. This prediction may then be compared to the observations. If the idea of inflation is correct, theory and observation should agree at some level. In view of this, we will discuss one very important observation that may allow the idea to be tested.

The irregularities in the density of the universe after inflation would have affected its temperature. The denser regions of the universe would have had a slightly higher temperature than the less dense regions. These differences in temperature between the high- and low-density regions survived as the universe expanded. They were still present when the matter and radiation stopped interacting directly with one another at the onset of the matter era. The cosmic radiation from the high-density regions would have had a slightly higher temperature than the average at that time. Conversely, radiation from the low-density areas would have been slightly cooler.

As we have emphasized previously, this radiation has remained essentially undisturbed since the end of the big bang. Only its wavelength has altered, because the universe has continued to expand. The temperature differences should still be present today in the cosmic microwave background. In other words, the radiation that comes from one part of the universe should have a slightly different temperature than radiation that arrives from another part.

These temperature irregularities, or 'ripples', are tiny. This makes them extremely difficult to detect. Typically, the temperature varies by less than one thousandth of a percent as we look out in different directions. It is rather like looking at a large brick wall from a great distance. The wall appears to be perfectly smooth to the naked eye. The individual pieces of stone that make up the complete building are too small to be detected. We can see them only by looking through a powerful pair of binoculars.

Until recently, cosmologists were rather like the observers who did not have access to such binoculars. They could see the wall – that is, the cosmic radiation – but lacked the necessary technology to see the individual bricks. They knew that the ripples in the cosmic radiation had to exist if galaxies had formed in the way that we described in Chapter 7, and researchers designed many experiments in an attempt to observe these ripples.

Despite much effort by cosmologists, the irregularities eluded detection throughout the 1970s and 1980s. By the close of the 1980s, many of these scientists were rather worried by this lack of positive results. Indeed, belief in the entire big bang picture was in danger of collapsing, and this would have forced cosmologists back to the drawing board. It was at this critical time that a satellite experiment was launched into space. This satellite, known as the Cosmic Background Explorer, or COBE, carried three experiments specifically designed to investigate the nature of the cosmic radiation.

One of the COBE experiments was dedicated to the detection of the predicted temperature fluctuations. In April 1992 the discovery of these fluctuations was announced and made headline news around the world. The experiment had measured a difference in temperature of just thirty millionths of a degree. This observation is viewed by many as one of the most significant cosmological breakthroughs since the discovery of the cosmic radiation itself.

What are the implications of the COBE detection? Before the ripples had been observed, the earliest time scale that we could explore experimentally was the electroweak era. The universe would have been about 10^{-10} seconds old when this era drew to a close. However, particle accelerators can provide only a very narrow window to this era. Earlier times than this cannot be investigated by terrestrial experiments. As we go back in time, the next significant event was the end of the Grand Unified era. This corresponds to a much earlier time and a considerably higher energy scale and construction of a machine capable of probing such a scale is clearly impossible. It would seem that the Grand Unified era must lie beyond the reach of our experiments.

The COBE detection reverses this pessimistic conclusion. COBE directly measures what the universe looked like at the era of decoupling.

The idea is that the conditions at this time were strongly influenced by the conditions that applied during the Grand Unified era. Inflation theory predicts that the ripples in the cosmic radiation are due to tiny quantum fluctuations that arose when the universe was less than 10^{-35} seconds old. The precise nature of these ripples depends quite strongly on the physical processes that powered the inflationary expansion. Thus, with present-day cosmological observations we might be able to reconstruct a picture of the very early universe before the electroweak era.

It is possible that the superstring theory could be tested in this way as well. One of the problems with superstring theory is that it does not make many predictions that can be verified in the laboratory. Although it is an appealing idea, we do not yet have any direct evidence that it might be correct. For example, the theory predicts that the internal structure of the elementary particles should become relevant at the Planck scale (10^{-35} metres). However, it is highly unlikely that an experiment capable of probing this extremely short distance could ever be performed.

If inflation occurred shortly after the Planck era, the quantum fluctuations that resulted in the cosmic ripples would have been determined by the superstring theory, in which case the ripples should contain some information about the theory. These same quantum fluctuations would ultimately have caused galaxies and clusters of galaxies to form after the matter and radiation decoupled at the end of the big bang. The distribution of galaxies in the universe today should also depend closely on the quantum fluctuations that arose in the very early universe.

It is possible that the large-scale structure of the universe may have been determined by the superstring theory at the Planck time. If so, we may be able to probe the smallest objects that exist – the superstrings – by looking at the largest objects in the universe that extend over hundreds of thousands of light years. Ultimately, the universe itself might provide the laboratory we need for testing the theory of everything!

In this chapter we have seen how the problems of the standard big bang model are resolved if the universe underwent a period of very rapid expansion in its distant past. In particular, the formation of galaxies may proceed due to tiny quantum fluctuations that existed in the density of matter. Cosmologists now realise that there are many different ways of causing the universe to inflate some time before the first 10^{-35} seconds. In the following chapter, we will consider some further consequences of an inflationary expansion during this era.

10

The Eternal Universe

The Planck time is the furthest back in time that we can go before quantum gravitational effects become significant. It has been suggested that inflation may have occurred when the universe was just 10^{-43} seconds old. This is precisely the era when superstring theory is supposed to be important.

More significantly, at least for our discussion, is that all the spatial dimensions of the universe would have had similar sizes. We discussed some of the theoretical arguments as to why there may be higher dimensions in Chapter 6. The existence of these extra dimensions has always been viewed as a problem. We know that the universe today contains only three large space dimensions, so why have the others remained so small that they cannot be seen? In other words, why is it that only three of the dimensions have grown to cosmological sizes? What is it that prevents the universe from having a different number of large dimensions?

Inflation, by its very nature, increases the size of the universe by a huge factor in a very small amount of time. We might therefore expect it to shed some light on the question of why three of the dimensions in the universe are so much larger than the others. The problem of small dimensions would be solved if only three of them were able to inflate to large sizes.

To explore this possibility further, let us now return to the question of initial conditions in the universe in Linde's chaotic inflationary scenario. Suppose that inflation had not yet occurred when the universe was just one Planck-time old. The Planck length, 10^{-35} metres, represents the maximum distance over which physical processes could have operated at that time. It does not, however, necessarily represent the actual size of the universe. The universe may have been much bigger than the Planck length. If so, it would have consisted of a collection of little regions that were approximately 10^{-35} metres wide. These regions were effectively disconnected from one another, because a light signal had not yet had sufficient time to travel from one region to another. In this sense, the conditions varied from region to region in a *chaotic* way. This picture of the universe before inflation is shown in Figure 10.1.

The physical conditions in each of the tiny regions would have been *uncorrelated*. There would have been no relationship between the conditions in one region and the conditions in a neighbouring one. The initial energy density – as determined by the potential energy of the ball in Figure 9.2 – would have varied from region to region.

How would this feature have affected the inflationary process? Linde realised that the chaotic nature of the initial conditions in the pre-inflationary universe would have had significant consequences. In our analogy, the ball would initially have been located nearer to the cliff in some regions than in others. The amount of inflation that occurs is

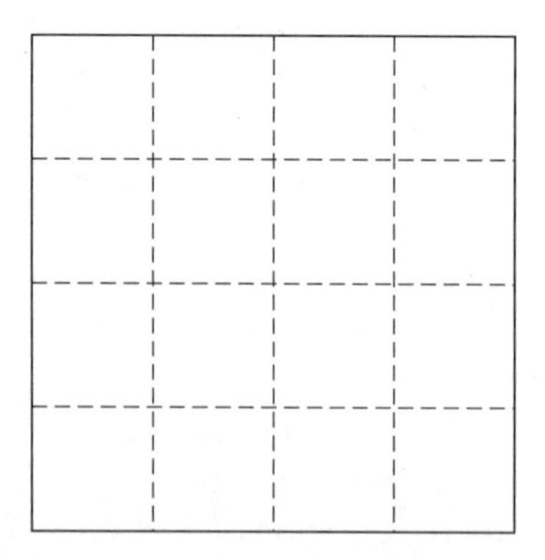

Figure 10.1. A representation of the chaotic inflationary universe at the Planck time before inflation has started. The universe may be viewed as a collection of Planck-sized regions. Conditions in these regions are uncorrelated. Some regions will inflate by a larger factor than others. The number of spatial dimensions that undergo inflation will also vary.

determined by the time it takes the ball to roll down the plateau and into the valley. More inflation is possible when the ball is farther to the right on the plateau since it takes longer to reach the valley. It follows that each Planck-sized region of the universe would have inflated by a *different* amount. Some regions would have grown to a huge size, whereas others may not have inflated at all.

Since these regions are disconnected from one another, we have no reason to suppose that the same number of dimensions inflated in each case. It is possible that only three of the dimensions grew in size in some regions and that the others remained static. Alternatively, a different number of dimensions may have inflated in other regions. In short, *all* possible combinations of large and small dimensions would have arisen. This implies that no underlying principle is required to ensure that only three dimensions grow to cosmological sizes. The whole universe may consist of many regions of different sizes each containing its own number of large dimensions.

A two-dimensional analogy will be helpful at this point in our discussion. Let us view Figure 10.1 as a two-dimensional elastic sheet. Suppose we were to stretch the elastic in such a way that each block expanded by a different amount and in different directions. This would represent the inflationary expansion of the universe. Some of the squares would expand more than others and would cover a greater surface area. Other regions would expand only in one direction and would resemble a thin line. The final result is shown in Figure 10.2.

Suppose we were then to place an ant onto one of these thin lines. This ant is aware only of the two dimensions of length and breadth that are associated with the elastic sheet. Since the second dimension in this region of the sheet is so small, the ant will be unable to detect it and will conclude that its universe contains just one dimension of space. If the elastic has been stretched sufficiently, the line on which the ant is balanced will be very long. It could be so long, in fact, that the ant will be unable to see beyond it to where the sheet extends out into the second dimension.

The idea is that we live in a region of the universe similar to that of the ant's. Only three of the space dimensions inflated. These are the only ones visible to us. The other dimensions are hidden from our view because they did not inflate and are too small to be detected. The region that we inhabit was also one of those that inflated by a huge amount. It extends well beyond our current horizon distance of ten billion light years. (Recall that this is the maximum distance that light

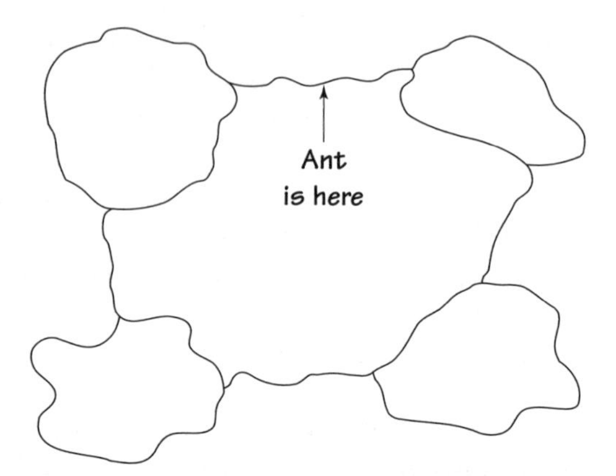

Figure 10.2. The two-dimensional elastic sheet has been stretched by different amounts in different directions. This corresponds to the universe after inflation. In some regions, only one of the dimensions associated with the sheet will be large. If an ant is placed into this region, it will conclude that its universe is a one-dimensional line. If this region is very large, the other areas of the sheet will lie beyond the ant's horizon and will be hidden from its view.

could have travelled since the end of inflation.) We cannot see the other regions that contain a different number of large dimensions, since light has not had enough time to move out of our region and into another one.

The part of the universe that we occupy may be considerably larger than the region of space that we can observe. The current horizon distance of ten billion light years corresponds to roughly 10^{26} metres. In some models of inflation, our region may extend out for a further 10^{10000} metres before it meets another region, with a different number of dimensions. This is a huge number; it is a ten followed by ten thousand zeros! The inflationary scenario suggests that the region of the universe we observe is just a *tiny* fraction of the total amount of space that actually exists. If we thought that the observable part of the universe was large, inflation predicts that the entire universe is significantly larger (see Figure 10.3).

Another remarkable feature of inflation is that it may never completely end once it has started. That is, inflation may be an *eternal* process. In 1986, Linde discovered that chaotic inflation could last indefinitely. We may understand why this might be the case by considering what happens during inflation inside each of the separate regions of Figure 10.1. As we have seen, each of these regions has a certain energy density at the onset of inflation that may be represented by the

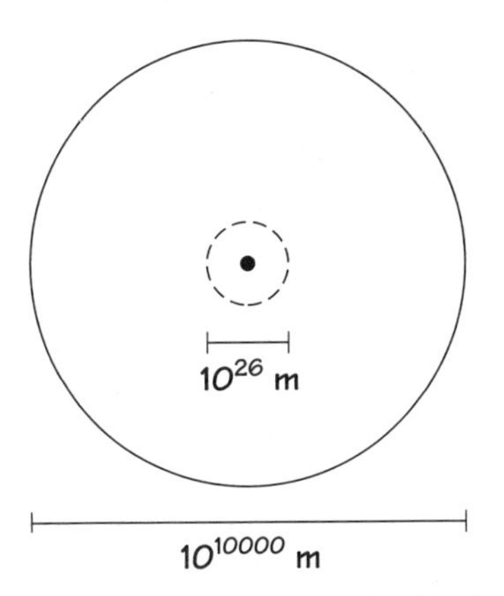

Figure 10.3. The observable universe, as represented by the dashed circle, may be just a tiny fraction of a much larger region of space. In some models of inflation this may extend for 10^{10000} metres. The figure is not drawn to scale.

height of the ball in Figure 9.2. Inflation arises when the ball is on the plateau and ends when it falls into the valley. Inflation is stronger, in the sense that the expansion rate is higher, when the ball is nearer to the cliff.

The motion of this ball along the plateau is determined by two factors. The ball has a natural tendency to roll down into the valley and thereby end inflation, but the role of quantum fluctuations must also be accounted for. In general, the quantum fluctuations are more pronounced when the ball is farther away from the valley. If the ball is sufficiently close to the cliff, they can dominate its motion.

The crucial feature of these fluctuations is that they can cause the ball to move either upwards or downwards along the plateau. In the vast majority of regions, the ball moves downwards and eventually reaches the valley. The standard big bang picture that we described in Chapter 7 will then apply in these regions after the inflationary expansion is halted. On the other hand, there are a small number of regions where the ball does indeed move upwards. Because these regions expand so rapidly compared to those where the ball moves downwards, they acquire a very large and ever-increasing volume. Inflation will never end in these regions.

In this picture, inflation need not stop in *all* regions of the universe once it has started. There are some regions where inflation proceeds indefinitely. These are surrounded by regions where inflation has stopped, and we inhabit one such region today.

Let us consider what happens to one of the regions that continues to inflate. We may view this region as a strip of elastic, as shown in Figure 10.4a. Consider two observers, *A* and *B*, who are placed on this strip and are initially separated by one Planck length. Let us suppose that the inflationary expansion causes the universe to *double* in size for each Planck time that elapses. Thus, the separation between *A* and *B* will double during this time interval, as indicated in Figure 10.4b.

Let us further suppose that *A* attempts to communicate with *B* by sending a light pulse when the two observers are separated by just one Planck length. This pulse travels at the speed of light, so it will have covered a distance of one Planck length after one Planck time has elapsed. It will reach the point originally occupied by *B* when it was sent out on its journey by *A*. This point is labelled *C* in Figure 10.4b.

By this time *B* will have moved two Planck lengths away from *A*. The light pulse from *A* will not reach *B*, and the latter will have no idea that the former is attempting to get in touch. What will happen after two Planck times? The light pulse will then have travelled two Planck

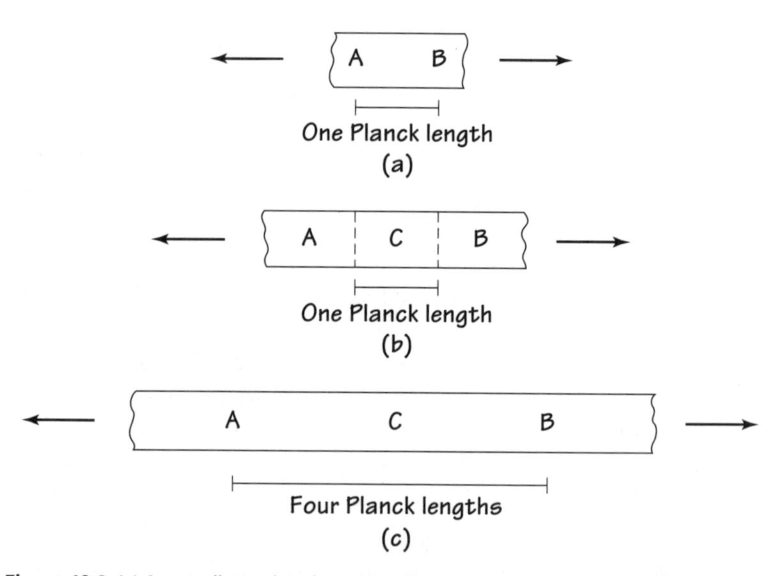

Figure 10.4. (*a*) A one-dimensional model where the universe is viewed as a piece of stretched elastic. At the onset of inflation two observers, *A* and *B*, are separated by a Planck length. *A* sends *B* a light pulse. (*b*) After one Planck time has elapsed, the light pulse has travelled one Planck length and is now located at *C*. The space (elastic) between the two observers has doubled in size, so *B* is two Planck lengths away from *A* and does not receive the pulse of light. (*c*) After two Planck times, the relative distance between the pulse and *B* has increased to two Planck lengths.

lengths, but *B* will be four Planck lengths away from *A*, as shown in Figure 10.4c. The relative separation between the pulse and *B* will be larger than it was after just one Planck time.

It is as if a greyhound is chasing after a hare that is running at twice its speed. The dog cannot match the pace of the hare, and the hare is able to evade capture. Likewise, the light pulse from *A* will *never* be able to catch up with *B* whilst the universe is inflating. Since communications cannot travel faster than light, *A* and *B* are effectively cut off from each other. This means that both observers will be isolated from any events that occur beyond a certain distance. In a sense, these observers will be surrounded by an horizon beyond which they cannot see.

This relative isolation is not specific to *A* and *B*. It applies to *all* regions of the inflating universe. Let us return to Figure 10.4b once more. During inflation, the space (elastic) between *A* and *B* is stretched and a new, Planck-sized region is formed around the point *C*. An observer located at this point will also be cut off from events that are occurring more than one Planck length away. Thus, a new isolated, inflating region will have appeared between *A* and *B* after one Planck time.

As the inflation of the original region of Figure 10.4a proceeds, it effectively splits into more and more of these isolated mini-regions. After one Planck time, there are two separate regions; the number increases to four after two Planck times, and so on. The total number of separate regions generated out of the original region doubles once every Planck time.

The expansion and future behaviour of these mini-regions is determined by the direction in which the quantum fluctuations act on the ball in Figure 9.2. In those regions where the ball moves downwards, inflation eventually ends, as we just discussed. In those cases where it is pushed upwards, inflation will continue indefinitely, thereby producing yet more mini-regions, and so on. This is important, because it implies that the production of new regions of inflating space is a never-ending process.

The eternity of inflation can in principle explain why our part of the universe looks the way that it does. There is a fundamental observation about the universe that we have not yet discussed. It is the fact that the universe is capable of supporting life. Although this observation may seem rather obvious, it nevertheless places severe restrictions on the structure of the universe.

For example, let us return to the question regarding the number of large dimensions in the universe. The inflationary picture implies that different regions of the universe will have different numbers of large dimensions, as depicted in Figure 10.2. Yet why is it that *our region* of the universe has three large dimensions rather than any other number?

The answer to this question is that we could not exist in a region of the universe where there existed a different number of large dimensions. Our existence depends crucially on the fact that we are located on a planet that orbits a star in just the right way. The existence of life requires the planet to remain a certain distance from the sun as it completes its orbit every year. If the Earth was positioned slightly nearer to or farther away from the sun, the temperature of the planet would be too hot or too cold to support life. The theory of planetary orbits implies that the Earth's orbit can be stable only when there are three large space dimensions. It would be unstable otherwise, and we would almost certainly not be able to survive.

The hydrogen atom could not exist if there were a different number of large dimensions. Hydrogen is a crucial component of water, and water is essential for life. We could not possibly exist without hydrogen, so our region of the universe must contain precisely three large dimensions.

A related question regarding the formation of elements such as carbon that are so crucial for life must also be considered. We have seen that carbon could not have been produced during the big bang because of the relative stability of the helium nucleus. It must have been generated in the centres of stars during the process of nuclear fusion. This process took billions of years to complete. A universe capable of supporting life must be able to survive for this length of time.

Furthermore, the stars themselves would not have been able to form if the universe had not been sufficiently close to the line *D* of Figure 8.1. If the universe had expanded too quickly, the reduction in the density of matter would have been too rapid. The matter in the universe would not have been able to clump together into galaxies and stars. Similarly, if the universe lay too far below the line *D*, it would have recollapsed before the carbon could be synthesized. In either case, life as we know it would probably not have been possible. We deduce by a process of elimination that the universe must lie very close to the line *D* if galaxies, stars and human beings are to exist.

We have seen that the inflationary universe generates many different regions of varying sizes with varying numbers of large dimensions. Our existence necessarily restricts us to a region where three large space dimensions are present and where gravity and the other forces of nature have just the right properties for life to exist. Although these restrictions on our region of the universe are very strong, the eternal nature of inflation implies that a region satisfying these constraints will eventually be formed.

In conclusion, we have discussed some further consequences of the inflationary universe in this chapter. We have seen that inflation provides a natural mechanism for producing large spatial dimensions in the universe. Moreover, it is possible that inflation never ends once it has started, in the sense that there always exists a region of the universe that is undergoing a rapid expansion. Inflation may therefore be an eternal process.

In the final chapter of this book we will consider to what extent the universe had a definite origin. Before doing so, we will discuss the class of objects known as 'black holes'. The internal structure of black holes may be relevant to the birth of our universe.

11

Black Holes

Consider what happens when you observe a particular object, such as a page of this book. Light, either from the sun or a lamp, is continually being reflected off the page, and some of this light will enter your eyes. This will stimulate the retina, thereby causing a signal to be sent via the optic nerve to your brain. The brain then deciphers this signal and 'reads' the words that are printed. The key point is that light has to be *reflected* if you are to see the words. In effect, it must *escape* from the surface of the page.

Suppose that we are located on the surface of a star and, using a cannon, fire a tennis ball upwards. The distance travelled by the ball is determined by the speed it has when it is released. If the ball is moving sufficiently fast, it can escape completely from the star's influence. In that case, it need never fall back to the surface.

What would happen if the star began to collapse? Its density would gradually increase, and the force of gravity near its surface would become stronger. This means that the ball would have to be released with a greater speed than before in order to escape. If the collapse were to proceed unhindered, the density and gravitational pull of the star would soon become extremely high. At some stage in the collapse the ball would be unable to escape from the star unless it travelled at speeds greater than the speed of light. Since nothing can travel faster than light, the ball would *never* be able to escape once this point had been reached. It would always return to the surface regardless of how fast it was fired from the cannon.

We draw similar conclusions when we consider the behaviour of light emitted from the star's surface. Before the star has collapsed, this light is able to escape and travel outwards through space. It could reach an observer who is many light years away from the star, and the star could be seen, at least in principle. Escape also becomes impossible for the radiation once the gravitational force around the star has become sufficiently strong. Hence, no more light is able to reach the distant observer, and the star effectively disappears from view.

At this stage the star is said to have formed a *black hole*. To view the star subsequently, our observer would have to reflect some light off its surface. Yet any light that makes contact with the surface from now on will be unable to overcome the intense gravitational force that is in operation. It will be unable to travel back into the eyes of the observer. The star can no longer be seen directly, and it is effectively black. Any matter that gets too close to the star will be unable to escape for the same reasons. If our observer were to throw an object, such as a tennis ball, at the star, the ball would become separated from the rest of the universe. As far as the observer was concerned, the ball would appear to have fallen into a cosmic hole.

If we are going to view them as more than just a subject for science-fiction stories, we need to understand how black holes might form in reality. In the preceding discussion we touched on the possibility that a star might collapse into a black hole. It will therefore be helpful to reconsider what happens to a star when it has exhausted its supply of nuclear fuel.

We saw in Chapter 7 that a star with a mass that is less than one and a half times the mass of the sun eventually collapses into a white dwarf. A white dwarf is stable against further collapse because the inward force of gravity is balanced by an outward pressure arising from the electrons. The force of gravity in a more massive star is strong enough to overcome this electron pressure. Such a star undergoes further collapse. In some instances, the core becomes dominated by neutrons. The fermionic nature of the neutrons produces a similar pressure to that of the electrons in white dwarfs.

If the mass of the star is not too high, the collapse of the core is halted by the pressure of the neutrons. The resulting object is called a *neutron star*. A neutron star has a radius of about fifteen kilometres and is extremely dense.

Can gravity ever be strong enough to overcome the neutron pressure? We might expect this to be possible if the star is very massive. In this case, the neutrons are unable to prevent further collapse. The neutron pressure will break down if the mass of the star is greater than a few times the mass of the sun. Such a star continues to shrink as the density and force of gravity increase. There is now nothing to prevent the collapse, and the star rapidly forms into a black hole.

What would we see if we were to watch a massive star collapse into a black hole? The answer to this question depends crucially on where we happen to be at the time of collapse. If we were some distance away from the star, we would see an entirely different set of events from those seen by someone who was on the surface.

Suppose a university professor wants to find out more about the collapsing star. Quite wisely, the professor thinks it might be a little too dangerous to remain on the star's surface as the collapse proceeds. She prefers to remain a safe distance away and encourages a research student to stand on the surface. The student, only too anxious to please his supervisor, readily agrees.

The student promises to communicate with his supervisor by sending out some form of electromagnetic waves. This could be light or radio waves, for example. When the signals begin to reach her, the supervisor measures the time that elapses between the arrival of successive peaks in the waves.

All goes well initially, and the peaks arrive at regular intervals. As time proceeds, the collapse gathers pace, and the density of the star becomes progressively higher, making it more difficult for light to escape from the surface of the star. The electromagnetic radiation sent out by the student has to overcome this extra resistance to reach the supervisor, and it loses energy in the process. This decrease in the energy of the radiation results in an increase in its wavelength, so the peaks in the waves begin to reach the supervisor less frequently than before.

Eventually, the collapse proceeds so far that no more electromagnetic radiation can escape to the supervisor. The professor loses contact with her student forever. This is the point at which the star forms into a black hole.

What does the student see at this time? He continues to send out signals at regular intervals and does not think anything unusual is happening. As the collapse proceeds the student does not feel anything special when the light he is sending ultimately becomes trapped by the star's gravity. In particular, he cannot tell that his signals are no longer getting through to his supervisor. As far as the student is concerned, he is still sending radiation to his supervisor as regularly as before.

What is the ultimate fate of the student? According to Einstein's theory, the strength of gravity is so strong when the light becomes trapped that it overcomes any further resistance to collapse. The theory predicts that the star will continue to shrink until it effectively vanishes into a single point of infinite density. This point is referred to as a *singularity*. In many ways, this singularity is similar to the big bang singularity that we discussed towards the end of Chapter 4. The laws of physics as we know them break down at this point. In short, the outlook is bleak for the student, who is doomed to destruction once the star has formed a black hole.

Thus far, we have discussed black holes only in rather qualitative terms. If we are going to proceed further, we will need a more precise description. Consider a particle of light – a photon – that is located some distance away from the singularity of the black hole. The gravitational influence of this singularity weakens as we move farther away from it. The photon will be able to escape the influence of the singularity if the relative separation between the two is sufficiently great. If, instead, the photon is initially very close to the singularity, it will inevitably get drawn towards it by the force of gravity.

This suggests the existence of a limiting distance around the singularity where the photon is only just trapped. This is known as the *event horizon*. We may view this horizon as a kind of bubble that shields the singularity from the view of a distant observer. If the photon is located outside the event horizon, it can still reach the observer. If it is located within the horizon, it can never get out. Indeed, in that case the photon will inevitably fall into the singularity and will be forever lost to the outside world.

In some sense, the event horizon may be viewed as the boundary of the black hole. Since photons located within this region cannot get out to a distant observer, all events that occur inside are hidden from our view. We may think of a black hole as the region of space that is contained within the event horizon.

This picture of a black hole is shown schematically in Figure 11.1. The singularity and event horizon are depicted by the dot and the circle, respectively. The interior of the black hole corresponds to the region contained within this circle. The paths of two photons, A and B, are shown by the solid lines. Photon A remains some distance away from the black hole. Although it is deflected by the gravitational pull

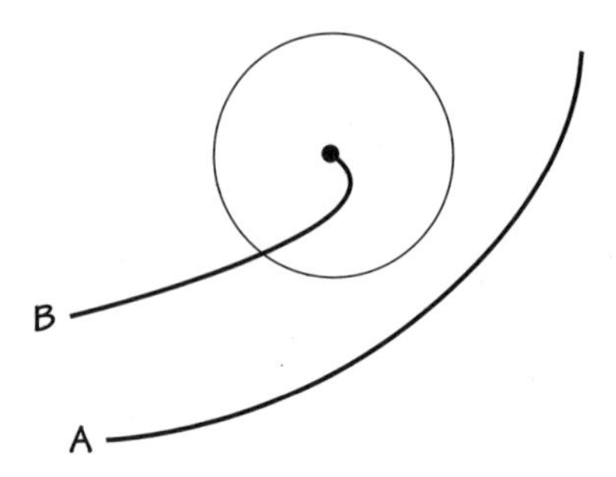

Figure 11.1. A schematic diagram of a black hole. The black dot represents the singularity. This is surrounded by an event horizon, as depicted by the circle. Photon *A* is deflected by the black hole but does not fall into it. Photon *B* gets too close and does cross the event horizon.

of the black hole, this photon does not cross the event horizon. It is free to continue on its journey through the universe. Photon *B* is not so fortunate. It is shown passing through the event horizon. It then becomes trapped by the black hole and soon falls into the singularity.

In the 1960s people began to seriously consider the possibility that black holes might exist in the universe. Until that time, it had not been clear to researchers that the very large densities required for the formation of a black hole could be obtained in practice. The situation changed somewhat with the detection of the first neutron star in 1967 by astronomers based at Cambridge, England. The typical size of a neutron star is only ten times greater than that of a black hole with an identical mass. This observation was important because it showed that objects with very high densities really did exist in the universe.

The discovery that Einstein's theory allowed for the existence of black holes was first made by the German astronomer Karl Schwarzschild in 1917. This was shortly after Einstein had published his general theory of relativity. The relevance of Schwarzschild's result to the real universe did not become apparent until the late 1960s. Schwarzschild had had to make many simplifying assumptions to derive his result. These were considered unphysical, and the general view was that realistic black holes would be far more complicated.

A number of significant theoretical breakthroughs made by researchers during the late 1960s and early 1970s changed this view. Let us return to the professor whose student has just been lost to the black hole. Does the supervisor have any way of showing that the student fell into the black hole? More generally, can the supervisor obtain any information regarding the nature of the matter that formed the black hole in the first place?

The answer to this question has far-reaching implications. Naively, we might think that the desired information would be available, at least in principle. It might be expected that the event horizon would depend in some way on the type of matter that collapsed. For example, consider the differences between a soccer ball and an American football. The former is spherical, whereas the latter is an ellipsoid. The soccer ball has a greater amount of symmetry. We can rotate it about any axis, and it will still look the same to us. We say that the soccer ball is spherically symmetric. This is not the case for the American football. Its appearance depends on the angle at which we happen to be holding it.

Suppose we were to squash both the soccer ball and the American football so much that they both collapsed into separate black holes. Since the two balls initially have very different shapes, we would expect the corresponding event horizons of the newly formed black holes to differ as well. Remarkably, this is not the case. Both event horizons will have a similar shape. It will not be possible to tell which black hole formed out of the soccer ball and which formed from the American football.

The reason for this is that any collapsing object radiates away some of its energy in the form of gravity waves. Gravity waves, which travel at the speed of light, may be thought of as ripples in the fabric of space-time. In many ways they resemble ripples on the surface of a pond. Any initial irregularities in the collapsing object are smoothed out and radiated away by the gravity waves. As a result, the final state of the black hole is very symmetric, regardless of how irregular the matter may have been at the onset of collapse.

The final state of any black hole that forms out of ordinary matter depends on only three fundamental quantities. A black hole has a mass and may also have electric charge and spin. These are determined by the mass, electric charge and spin of the original object that forms the black hole. A black hole has no other defining features. All other quantities, such as its size, are uniquely determined by these three parameters. This is very important because it implies that the relatively simple black holes, such as the one discussed originally by Schwarzschild, may be quite realistic after all. Moreover, when two different objects with the same mass, charge and spin collapse, the resulting black holes will be indistinguishable to an outside observer.

This characteristic of black holes may be summarized by saying that black holes have no 'hair'. If we were to observe two completely bald people, we would not be able to deduce their original hair styles. Regardless of the initial style, the final result is the same for each person. Likewise, the final state of the black hole is the same regardless of the initial appearance of the matter once the mass, charge and spin have been determined.

A natural question to ask at this stage in our discussion is whether the fundamental laws of physics still hold in the very extreme environment near a black hole. One important law is known as the *second law of thermodynamics*. This law determines how an isolated system gradually becomes more and more disordered over time, and it has profound consequences for our understanding of black holes. In view of this, we will now make a small detour and consider the consequences of this law further.

We may understand the principle behind the second law of thermodynamics by first considering another fundamental law that is more familiar. This is the law of energy conservation. This states that *the total energy of an isolated system remains unchanged with time*. We have already referred to the principle of energy conservation in earlier chapters, but it will be helpful to discuss it in more detail.

Although the amount of energy in an isolated system can neither decrease nor increase, the precise form that the energy takes may alter over time. To illustrate this point, let us consider the valley shown in Figure 11.2a. If we were to release a ball from rest some distance up the hill, it would begin to fall towards the bottom of the valley under the force of gravity. Initially, all of the energy associated with the ball will be stored in the form of potential energy. As the ball rolls down

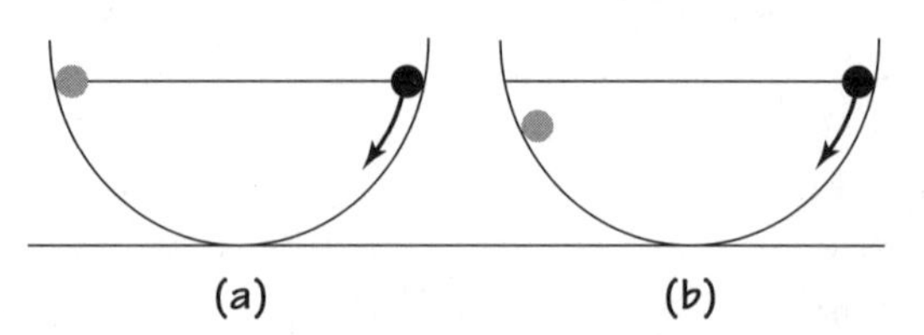

Figure 11.2. *(a)* In an ideal world, a ball released from rest will roll down and reach the same height that it started with. *(b)* In reality, the effects of friction mean that the maximum height attained by the ball is somewhat lower each time it rolls back up the hill. Although energy is conserved throughout, the amount that is available for moving the ball steadily decreases.

the hill, it gathers speed, and its potential energy is converted into the kinetic energy of motion.

If there is no friction acting on the ball, all of its potential energy will have been converted into kinetic energy by the time it reaches the bottom of the valley. This kinetic energy would then drive the ball up the other side of the hill. The ball slows down as it climbs the slope, so its kinetic energy is gradually converted back into potential energy. Eventually, the ball will instantaneously stop when all of its kinetic energy has been converted. The height attained by the ball when this occurs will be identical to the height it had when it was initially released.

This process is then repeated indefinitely as the ball rolls from one side of the valley to the other. The total energy of the ball – as given by the sum of its kinetic and potential energies – remains constant, although the *type* of energy associated with the ball does vary from point to point.

Like the law of energy conservation, the second law of thermodynamics is also concerned with energy, but it deals with what can be done with a given amount of energy. It provides a more accurate description of what really happens to the ball as it rolls back and forth. We know from experience that the maximum height attained by the ball after each oscillation will gradually decrease until it eventually comes to rest at the bottom of the valley.

The ball loses its energy to outside influences such as friction. Over time, less and less energy is available to the ball to enable it to climb the hill. The second law takes account of this. More generally, it states that the *amount* of energy that is available for doing useful work inevitably decreases as time progresses.

In the preceding example, the effects of friction due to air resistance and the contact of the ball with the ground mean that not all of the ball's potential energy is converted into kinetic energy. Some of the potential energy is used in overcoming these frictional forces. Air molecules are continually bouncing off the ball, and this introduces a viscosity. The air molecules gain energy from the ball, slowing it down. Thus the ball has lost some of its original energy by the time it gets to the other side of the valley, and it fails to reach the same height as before, as shown in Figure 11.2b. This transfer of kinetic energy away from the ball proceeds during subsequent cycles until the ball loses all of its potential energy to the air molecules.

It is important to emphasize that the total energy of the system – which now constitutes the ball and the atmosphere – remains constant. However, the amount of energy available to the ball gradually diminishes. This decrease in the amount of useful energy may be measured in terms of the disorder in the system. The system is initially in a highly ordered state, because all of its energy is associated with the ball. The final state of the system is very disordered, because its energy has been distributed amongst the countless air molecules. In other words, the system gradually becomes more and more *disordered* over time. We may therefore associate a loss of useful energy with an increase in disorder.

The amount of disorder in a system is usually expressed in terms of a quantity known as *entropy*. The higher the entropy of a system, the more disordered it is. Conversely, a more ordered state has a lower entropy. In general, the second law of thermodynamics tells us that *the entropy of an isolated system will increase with time.*

The tragic fairy tale of Humpty Dumpty provides us with an example of the second law of thermodynamics in action. When Humpty is perched on the wall, he is 'together' and in a well-ordered state of low entropy. When he falls he breaks into numerous pieces and is clearly in a more disordered state than beforehand. Thus, Humpty's entropy has increased. Spontaneously putting him back together again would leave him in a more ordered state and would decrease his entropy, but this is forbidden by the second law.

We are now in a position to see how the second law of thermodynamics is related to the nature of black holes. Let us begin by considering what would happen if a complex object were to be thrown into a black hole. By their nature, black holes suck in any object that gets too close, whether it be a star, radiation or even an astronaut! If two black holes collide, they will merge into an even bigger black hole. According to the no-hair property discussed earlier, only the mass, charge or spin of a black hole can be altered if something falls into it. For example, since mass and energy are equivalent, the mass of a black hole will increase when it gains energy, say, in the form of radiation.

Suppose we were to hold two identical cups just outside the event horizon of a black hole. Together these cups have a certain amount of entropy. If we were then to break one of them, the combined entropy of the two cups would increase for the reasons we discussed earlier. What would happen to the entropy if we were then to throw the pieces of the broken cup into the black hole?

According to the no-hair property, the only effect the pieces of the broken cup would have on the black hole would be to increase its mass slightly. (We are assuming, for simplicity, that the pieces of the broken cup do not carry any net electric charge or spin.) The pieces of the broken cup will disappear from view behind the event horizon, and only the single unbroken cup will remain. The cup that is intact has much less entropy than the broken one, since the latter is very disordered. A certain amount of entropy will therefore have been lost inside the black hole. A distant observer will register a *decrease* in entropy in the universe outside the black hole, and this is in direct violation of the second law of thermodynamics.

This argument seems to suggest that the second law may not apply near a black hole. The essential question that needs to be addressed is whether this paradox can be resolved. In other words, is there some crucial feature exhibited by black holes that is capable of incorporating the second law of thermodynamics into the picture? This was the burning question being considered in the early 1970s, and many researchers were quite puzzled. A young student at Princeton University, Jacob Bekenstein, suggested how the problem might be resolved.

Bekenstein had been inspired by a previous discovery of Stephen Hawking. Hawking had considered what would happen to the event horizon of a black hole when an object, such as a broken cup, falls into it. The cup has energy in the form of mass, so the mass of the black hole will increase. A higher mass implies that light will become trapped by the singularity over a greater range than before. Consequently, the event horizon of the black hole will be located farther away from the singularity. Since the size of a black hole is determined by the size of its event horizon, the black hole will increase in both size and mass when the pieces of the broken cup are thrown into it.

This implies that the event horizon of the black hole will cover a greater *area* than before, for the same reason that the area of a sphere increases as its radius increases. Hawking proved rigorously that the area of the event horizon *always increases* in any physical process involving the black hole.

A link between this result and the second law of thermodynamics may now be made. As we have seen, the second law implies that the entropy (that is, the amount of disorder) of a system always increases. According to Hawking, so too does the area of a black hole. It is natural to wonder whether the area of the event horizon might be directly related to the entropy of a black hole.

This was Bekenstein's idea. He suggested that the size of the event horizon should indeed be viewed as the entropy associated with the black hole. In this case, the overall entropy can increase when the broken cup falls into the black hole, as required. The second law of thermodynamics need not be violated if the area of the black hole grows in such a way that the *total* entropy of the one remaining cup and the black hole afterwards is *greater* than the combined entropy of the black hole and two unbroken cups beforehand.

There is a problem with this idea for the following reason. Consider what would happen if we were to leave a freshly cooked pie in a sealed room. Initially, the pie will be considerably hotter than its surroundings. Over time, it will cool down as its heat is distributed around the room. Conservation of energy then implies that the room will heat up, since the total energy associated with the pie and the room must remain constant. This exchange of heat continues until both the pie and its surroundings reach the same temperature. At this point, the pie and the room are said to be in a state of thermal equilibrium.

What happens to the combined entropy of the pie and room during this process? Initially, the system is well ordered, because all of the heat is concentrated in the contents of the pie. The entropy of the pie and its surroundings will be low. As the pie cools, the heat is distributed randomly about the room, and the system becomes highly disordered. The final amount of entropy is greater, as expected.

The key point in this discussion is that an increase in entropy is intimately related to the transfer of heat from a hot to a cold body. This implies that temperature and entropy must be related in some way, since heat is always associated with temperature. In particular, it implies that an object must have a certain temperature above absolute zero if it has entropy. This conclusion should apply to all objects, including black holes. A black hole will have a temperature if it has entropy.

Consider a single black hole with a certain temperature. If this black hole is hotter than its surroundings, it should radiate. However, the defining feature of a black hole is that nothing – not even electromagnetic radiation – can ever escape from its surface. This suggests that a black hole should never be able to radiate, and this is where the problem arises.

Hawking resolved this contradiction by accounting for quantum fluctuations that arise just outside the black hole's event horizon. He showed that these fluctuations allow the black hole to effectively radiate particles and radiation.

We will discuss Hawking's result in the remainder of this chapter. Let us begin by recalling our discussion from Chapter 5 regarding the nature of the vacuum. When quantum fluctuations are ignored, the vacuum, by definition, contains no matter or energy. It is therefore completely empty.

Quantum fluctuations modify this picture significantly. The energy of a particle can never be measured precisely. This principle applies to everything, including the vacuum. Hence, we can never verify that the vacuum is completely empty at all times, because it takes a finite amount of time to complete any measurement. Nothing prevents particle-antiparticle pairs from popping into existence out of the vacuum. These particles are known as virtual particles, because they annihilate each other before any measurement can be made. They can never be directly detected.

A similar uncertainty produces particle-antiparticle pairs from the empty space near the event horizon of a black hole. What happens to these pairs once they have been created? There are four possible scenarios, as shown in Figure 11.3. The creation of the virtual particles is illustrated by the solid lines. The particles are denoted by the symbol p and the antiparticles by $\bar{p}$. In case A, the pair destroy each other before either particle has had time to fall into the black hole. In case B, both particles fall into the black hole and then annihilate. In cases C and D, only one of the particles crosses the event horizon. In the former case, the particle becomes trapped by the black hole, whilst the

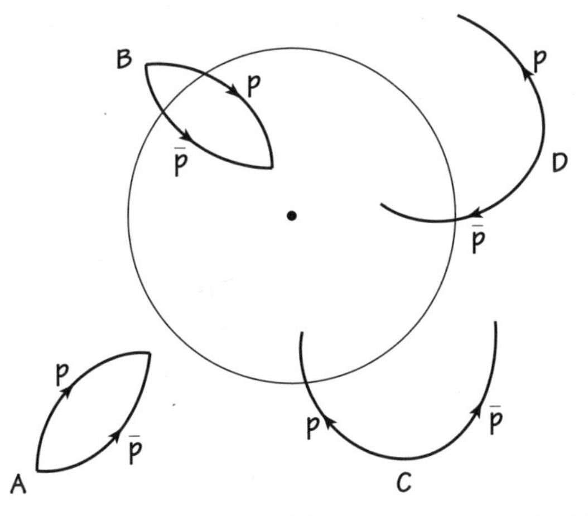

Figure 11.3. Quantum fluctuations result in the production of virtual particle-antiparticle pairs outside the event horizon of a black hole.

antiparticle remains outside. In the latter case, it is the particle that remains free and the antiparticle that becomes confined.

The production of these virtual particles around the black hole does not imply that energy is spontaneously created in the process. One of the virtual pair is created with a certain amount of positive energy, and the other is created with an equal amount of negative energy. This ensures that the combined energy of the pair is zero and that there is no overall change in energy.

We saw in Chapter 5 how the energy associated with the gravitational force is negative. In the circumstances that apply near the surface of the Earth, the positive energy associated with the mass of an object dominates any negative energy contributions from gravity. This is because the gravitational force of our planet is relatively weak. The total energy associated with the object remains positive.

This will not necessarily be the case if the same object is then moved near a black hole. The force of gravity around the black hole will be very strong, so a great deal of work must be done if we are to move the object away from the event horizon. The object will gain a huge amount of energy in the process, even if it is only moved by a small amount. Conversely, the object will lose a lot of positive energy (that is, gain a lot of negative energy) if it is carried towards, and ultimately into, the black hole.

The object will continue to gain negative energy after it has crossed the event horizon. Eventually, it will acquire so much negative energy that this will come to dominate any other forms of energy associated with the object. The object's overall energy will become negative. This argument also applies to any elementary particles that may fall into the black hole. A real particle can have an overall negative energy when inside a black hole.

Let us return to Figure 11.3. Hawking realized that the negative energy virtual particle *behaves as if it were a real particle* when it falls into the black hole. This means that the particle-antiparticle pair need not necessarily annihilate each other. In case *D*, for example, the particle with positive energy that remains outside the black hole can also become real. It can then move away from the event horizon and ultimately escape from the influence of the black hole altogether.

What happens in case *D* is that the black hole gains a little bit of negative energy as the virtual particle falls into it. This gain of negative energy is equivalent to a loss of positive energy. Since mass and energy are equivalent, the overall effect is that the black hole loses mass in the process.

What would an observer outside the event horizon see whilst this was occurring? When a particle with positive energy escapes as the black hole loses mass, it would seem to an observer as if the black hole had *emitted* the particle. The black hole would appear to have radiated away some of its energy. Since a radiating body has a temperature, the black hole may indeed have entropy.

The entropy associated with the black hole is determined by the area of the event horizon. A larger area implies a larger entropy and vice versa. The amount of entropy is directly related to the mass of the black hole, since a lower mass implies that a smaller area is covered by the event horizon. When the black hole emits the real particle, it loses mass, and its event horizon shrinks. Thus, its entropy decreases. This would seem to violate the second law of thermodynamics, but this is not the case. Although the entropy of the black hole does decrease, the total entropy of the black hole and the emitted particle is greater than that of the black hole before the emission. The overall entropy therefore increases.

To summarize thus far, black holes may have entropy and a non-zero temperature because the quantum fluctuations associated with the vacuum lead to the effective emission of real particles from the black hole. The creation of virtual particles around the black hole is a continuous process. The black hole acquires more and more negative energy particles and loses more of its mass. This process is referred to as *black hole evaporation*, since the black hole appears to a distant observer as if it were evaporating away its energy.

At what rate does the black hole evaporate? As the event horizon shrinks, the black hole acquires negative energy particles at an ever-increasing rate. Consequently, the evaporation of the black hole becomes more efficient as it proceeds. The evaporation causes the mass to decrease, and this in turn leads to a further reduction in the size of the event horizon. This results in an even faster evaporation, and so on. The temperature of the black hole will also increase as the black hole loses mass and the evaporation gathers pace. This follows since a hotter body radiates at a faster rate than a cooler one.

Evaporation is a relatively slow process until most of the black hole's mass has been lost. There is then a flurry of activity at the very end of the evaporation. To a distant observer, the black hole appears to be stable until it suddenly explodes in a burst of radiation and particles.

It is not presently clear what happens to the black hole during the final moments of its evaporation. Hawking had to make some simplifying assumptions when he performed his original calculations. Essentially, his analysis is valid until the black hole's mass has dropped to about one hundred thousandth of a gramme. Typically, a black hole of this mass is about a Planck length in size.

Two schools of thought exist as to what might happen once the black hole has reached this size. One view is that the black hole evaporates completely and simply disappears. The other option is that the black hole stops evaporating at this point and forms a stable remnant. Although both views have some support, there is currently no generally accepted answer to the question of which, if either, is correct.

The lifetime of a black hole is determined by the mass it has when it forms. A more massive black hole has a longer life expectancy because the evaporation process is less efficient than it would be for a smaller black hole. It also has a larger amount of energy that needs to be evaporated away. At the beginning of this chapter, we saw how very massive stars ultimately collapse into black holes. These black holes have an initial mass that is at least three times greater than that of the sun. Because this mass is very high, the evaporation is rather inefficient. It would take a black hole of this size at least 10^{64} years to evaporate completely. For comparison, the current age of the universe is only about 10^{10} years.

Does this mean that the evaporation of a black hole could never be observed? If so, such a process would remain an interesting theoretical speculation but would have no direct influence on the current state of the universe. We need to form black holes whose masses are so low that their evaporation would have been completed by the present time. The masses of these black holes must be significantly lower than the typical masses of stars.

How might such black holes arise? It is possible that they were produced in the big bang just after inflation had ended. We saw in Chapter 10 how the very rapid expansion associated with inflation allows us to view the inflating universe as a collection of tiny, isolated mini-regions. This picture also applies to the universe immediately after inflation. The horizon distance at that time determined the size of each region. We have further discussed how quantum fluctuations would inevitably have been present during inflation. These perturbations would have led to density perturbations in the postinflationary universe. In a small number of the mini-regions, the extent of these fluctuations would have been comparable to that of the horizon distance. This would have caused such regions to collapse into tiny black holes.

These objects are called *primordial black holes*. They can have a mass as small as a few grammes. The key point is that these black holes have relatively short lifetimes and will have already evaporated by now. Black holes that formed during the big bang with a mass of 10^{15} grammes should be evaporating at the present era.

This concludes our discussion on the properties of black holes. We briefly alluded to the fact that the singularity inside the black hole is similar to that associated with the big bang. In the final chapter, we will apply what we have discussed thus far in this book to explore such a connection further. This will lead us to the birth of the universe.

12

The Birth of the Universe

Our journey has been remarkably successful. We have followed the development of the universe from its earliest moments through to the present. Our picture is that the universe has been expanding ever since it came into existence some ten billion years ago. However, Einstein's theory of general relativity breaks down at extremely early times when the universe was very small. The critical time scale over which the theory does not apply is the Planck time, corresponding to 10^{-43} seconds. We cannot employ Einstein's theory to determine the nature of the universe before this time.

The question we will address in this concluding chapter is whether we can travel back beyond the Planck time. Are we able to complete the story and cross this final barrier, or does the breakdown of Einstein's theory at this point represent a fundamental limit to our understanding?

If we are going to discuss the origin of the universe, we must first extend Einstein's theory in some suitable way. How might we proceed to modify the theory? We may answer this question by appealing to quantum effects. We have seen in the preceding chapters how these effects become significant on very small scales. The reason why Einstein's theory breaks down before the Planck time is that it fails to take into account the quantum fluctuations that are inherently present in any physical process involving gravity.

These uncertainties are not important on large scales and can be ignored. This is why general relativity provides a good description of the universe over very large distances. However, the quantum fluctuations associated with gravity dominate at the Planck time, and Einstein's theory as it stands becomes unsuitable. The key point is that general relativity and quantum theory must both have played an important role before the Planck time. These two theories must be combined in some way before the origin of the universe can be discussed. We need to apply the ideas that lie behind quantum mechanics to the universe as a whole.

We are searching for a theory that describes the origin of the universe. Let us begin by considering the concept that the universe was *created out of nothing*. In Chapter 5 we saw how quantum fluctuations in the vacuum – that is, empty space – result in the spontaneous creation of particles and antiparticles. The idea is that a similar quantum fluctuation could also result in the creation of an entire universe. The argument goes as follows: initially, there was nothing, but then a quantum fluctuation occurred and caused a tiny, fledgling universe to emerge from the emptiness. This universe subsequently inflated and grew into the complicated structure that we observe today.

An analogy will help us to understand how this might have occurred. A cowboy hat is shown in Figure 12.1. This hat has a rim at its base, and it is sharply peaked towards the centre. Notice how the peak of the hat dips down slightly at the very centre.

Suppose we were to place a small ball at rest into the bottom of this peak. In this analogy, the ball will represent the universe in some sense.

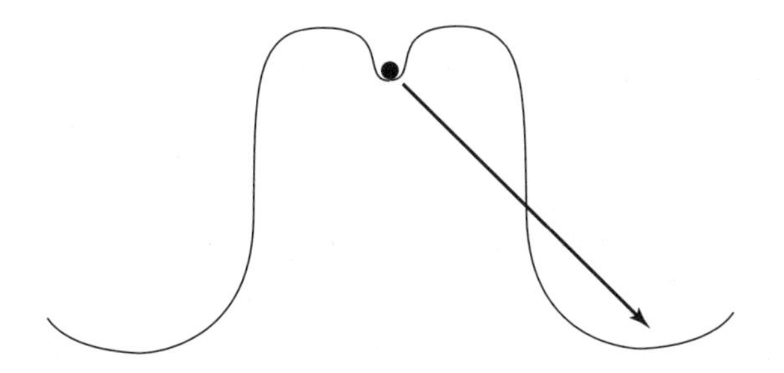

Figure 12.1. The creation of the universe from nothing may be viewed as the tunnelling of a small ball through a barrier. In the absence of quantum fluctuations, a ball that is initially at rest in the peak of the hat will remain there indefinitely. Quantum fluctuations in the position and energy of the ball imply that it will eventually be able to travel over the barrier and into the rim. The ball effectively tunnels through the hat. This event represents the creation of the universe.

We shall say that the universe does not exist when the ball is positioned in the peak of the hat. On the other hand, the universe does exist if the ball is located in the rim of the hat. Having defined the existence of the universe in this way, we may then say that it comes into existence when the ball moves from the peak of the hat down towards the rim. (The analogy is not precise because the ball clearly exists when it is on the peak, even though it is supposed to be representing a nonexistent universe. Nevertheless, the key idea is that the position of the ball on the hat determines the state of the universe.)

How can we arrange for the ball to travel over the barrier that separates it from the rim? We appear to be faced with a problem, because the ball is initially at rest and has no kinetic energy. This suggests that the ball must remain confined inside the peak indefinitely, because it does not have enough energy to start moving upwards over the bump. Quantum fluctuations change this conclusion. We have seen previously that the position of the ball cannot be determined precisely when quantum fluctuations are accounted for. Neither can its energy. Thus, the ball need not necessarily be confined precisely at the bottom of the peak. The uncertainty in its energy is random, and eventually this uncertainty becomes relatively large. This enables the ball to move over the barrier and into the rim of the hat.

This process is known as *tunnelling* because the ball effectively tunnels through the hat. Something similar could have happened to the universe. Quantum fluctuations allow it to tunnel into existence from a state of nothing. This description of the universe's birth as a tunnelling process has been advocated by the Russian school, and in particular, by Andrei Linde and Alexander Vilenkin.

We might wonder whether this description of a universe being created out of nothing is consistent with the other fundamental laws of physics, such as the conservation of energy. Energy conservation implies that we cannot get something for nothing, but the opposite appears to be true here. Indeed, we appear to be getting an entire universe for free! Does this mean that energy is not necessarily conserved after all?

The total amount of energy present in the universe must be zero if the universe was created out of nothing. We know that a certain amount of positive energy exists in the form of mass. There must also exist an equal amount of negative energy that precisely cancels out this positive energy. The required negative energy is provided by the force of gravity. The total negative energy associated with the gravitational effects of all the stars and galaxies in the universe can *exactly* cancel the positive energy associated with all of the mass. The total energy of the universe can be zero as required.

The same argument applies to electric charge. The total electric charge must have vanished before the universe came into existence because no elementary particles were present. If electric charge is always conserved, the total electric charge in the universe today must also vanish.

Remarkably, evidence suggests that this may indeed be the case. There are only two long-range forces operating in the universe. These are the force of gravity and the force of electromagnetism. The former operates between objects that carry mass, whereas the latter only affects electrically charged objects. The electromagnetic force is much stronger than the force of gravity, so it would dominate the dynamics of the universe if the universe contained an excess of positive or negative electric charge. However, observations indicate that the galaxies and stars interact with one another only through gravity; no electromagnetic effects are seen. We can conclude, therefore, that the combined electric charge of all the stars and galaxies in the universe adds up to zero. This implies that the total electric charge of the universe vanishes.

The possibility that the universe contains zero electric charge and energy has profound consequences. It suggests that everything we see in the universe adds up to nothing. In this sense, we are not really getting something for nothing as we previously thought.

This is only half the story, though. The idea that a quantum fluctuation can lead to the creation of the entire universe from nothing is very appealing, but a number of issues remain unresolved. It is far from clear how the quantum fluctuation arises in the first place. There is also the question of how space and time fit into this picture. The three essential ingredients in the universe are space, time and matter. Does the fluctuation occur in a preexisting space-time, or are space and time created by the fluctuation along with the matter? In other words, are space and time independent of the rest of the universe, or are they a fundamental part of it?

If space and time were indeed independent quantities, we could think of the creation of the universe as occurring at some definite moment in time. However, the ideas of general relativity have shown us how space, time and matter are intimately linked. It is natural to suppose, therefore, that all three should be created together. It would then follow that time would have a definite origin in this picture, because there would have been no such thing as time beforehand.

If we are to understand the nature of this origin, we must first establish what we mean by time. In short, time is something that we measure. For example, we define the year to be the time it takes the Earth to complete one orbit around the sun. Alternatively, we can define a unit of time as the period it takes the moon to complete one orbit around the Earth. When we were discussing Einstein's theory in Chapter 4, we defined a unit of time as the interval that elapsed between successive flashes from a lamp.

We also saw in Chapter 4 that we should view time as part of space-time. The entire universe then consists of space-time plus matter. It will be helpful to draw the space-time diagrams corresponding to some model universes. For example, let us consider a closed universe whose space is one-dimensional. This space may be viewed as a circle, and an observer in this universe would then be confined to live on this circle.

Let us further suppose that this model universe is static, so that it neither expands nor contracts. The size of this universe is uniquely determined by the radius of the circle, so the radius will stay fixed in time. We have shown the space-time diagram of this universe in Figure 12.2. Time flows upwards in this diagram, and space-time resembles a cylinder, since the size of the circle is the same for all times.

We know that our universe is expanding, so this picture must be extended if it is to represent the real world. In this one-dimensional

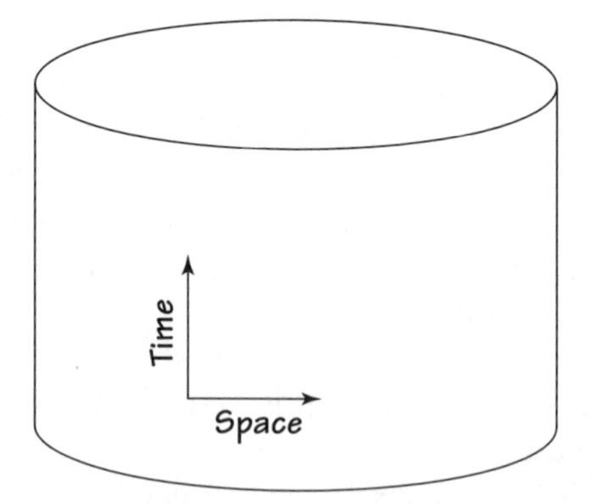

Figure 12.2. A closed, static universe with one space dimension may be viewed as a circle. Since this universe is not expanding, the radius of the circle remains fixed for all time. As this circle moves through the time dimension it traces out a two-dimensional surface in space-time. The space-time diagram of such a universe resembles a cylinder. Time flows vertically upwards, and the spatial dimension extends horizontally.

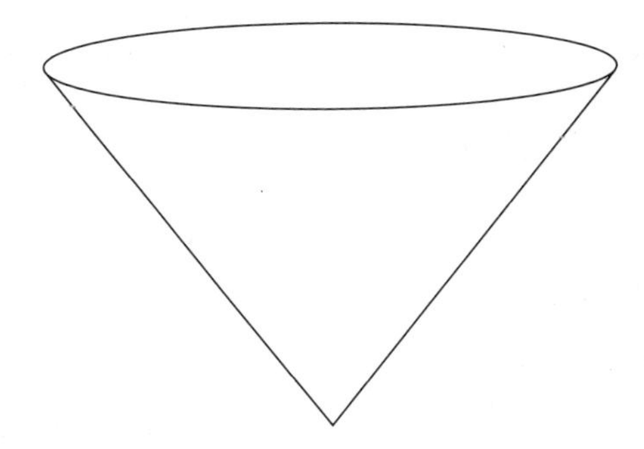

Figure 12.3. The space-time diagram for a one-dimensional, expanding universe. Space and time are associated with the horizontal and vertical directions, respectively. The spatial dimension of this universe is represented by a circle. Since this universe is expanding, the radius of the circle increases with time, thereby causing the space-time to resemble a cone. The tip of the cone corresponds to the point where space and time cease to exist. This point represents the big bang singularity.

analogy, an expansion of the universe would correspond to a progressive *increase* in the radius of the circle. In this sense, later times would correspond to larger radii. The space-time diagram of such an expanding universe would resemble a cone, and this is shown in Figure 12.3. As we extrapolate back in time in this picture, the radius of the circle (universe) decreases until it vanishes completely at the tip of the cone.

It is important to emphasize that in this picture, both space and time appear out of nothing along with the rest of the universe at the tip of the cone. The essential feature of this diagram is that the universe is described in terms of the *shape* of the space-time diagram. However, the quantity that we call time does not determine the shape of the space-time. It simply represents a convenient way of ordering events. In an expanding universe, it is natural to order events in terms of the size of the circle, with a zero radius corresponding to the zero of time. In this context, time is the means by which we measure the size of the universe.

A similar ordering of events occurs when we watch a film. We can consider a film in one of two ways. We can discuss it from the perspective of a viewer in the cinema, or alternatively, from the vantage point of the person in charge of the projector. The viewer sees a sequence of events and has an overwhelming impression that there is a flow of time as one frame of the film follows another. The projectionist has a different interpretation. As he inserts the film into the projector, he sees it as a self-contained entity. He does not identify a measure of time with the movie. Time becomes apparent only if each frame is viewed in succession. In a similar way, the flow of time in an expanding

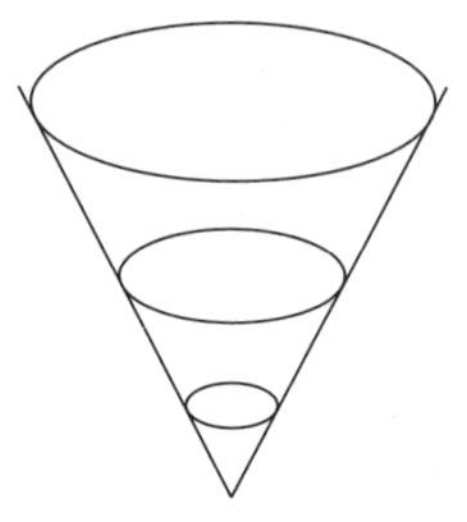

Figure 12.4. The cone representing the space-time of an expanding universe may be constructed by placing circles of progressively larger radii on top of one another. Each circle may be labelled in the same way that the exposures on a roll of cinema film are labelled. This labelling represents the flow of time in the universe. Time is an intrinsic property of the universe and is not external to it. In the figure on the right, time moves vertically upwards and space extends horizontally.

universe is apparent only to an observer who continuously measures the universe's size.

In a certain sense, therefore, we may think of the space-time in Figure 12.3 as a roll of cinema film. The film is made up of a series of separate frames that are joined together in a well-defined sequence. In a similar way, the space-time of Figure 12.3 may be viewed as a collection of slices that are positioned on top of one another. This correspondence is shown schematically in Figure 12.4. In this picture, each frame of the film represents a given slice of the universe. It is rather like considering individual slices in a loaf of bread. If we take slices of bread and place them on top of one another, we will recover the original loaf.

We have yet to consider how quantum fluctuations in space and time might affect the shape of these space-time diagrams. We have already argued that such effects cannot be ignored when the universe is extremely small. In Figure 12.3, the universe becomes smaller as we approach the tip of the cone. It is possible that the shape of the diagram in this region could become altered in some way by the quantum fluctuations.

Let us consider these effects further. It is important to emphasize that quantum fluctuations affect *all* measurements, including those of space and time. This implies that there will always be a limit to how accurately we can measure a given length scale. Such an uncertainty becomes relevant at distances of 10^{-35} metres, that is, at the Planck scale. This represents a fundamental limit to how precisely any length can be measured. We expect the quantum fluctuations in space to become relevant when the universe is smaller than this.

There is a similar limit to how precisely time can be measured. The simplest way to define a unit of time is to measure the interval it takes a swinging pendulum to complete one full oscillation. We can never do this precisely, because we cannot measure the position of the pendulum exactly. In particular, we cannot measure the precise instant when the pendulum reaches its point of maximum displacement. This leads to a fundamental uncertainty in the measured time it takes the pendulum to complete one full swing. The Planck time is the size of this uncertainty, and such an error is *always* present in any measurement of time. Naturally, when compared to the time scales that we measure in our everyday lives, this uncertainty is sufficiently small for it to be ignored. However, when the age of the universe was comparable to the Planck time, the effect was very important.

If space and time cannot be physically measured below the Planck limit, it is not clear that such quantities could have any physical meaning. The very concept of time may disappear just before we arrive at the tip of the cone, and at a certain level, the same might be true for space.

If this is the case, how might these quantum fluctuations in space and time affect a space-time diagram such as that shown in Figure 12.3? In this diagram, time runs vertically upwards, whereas space is confined to the horizontal direction. We lose any physical meaning of space and time below this scale due to the quantum fluctuations. The result is that we can no longer say that time runs upwards and that space extends horizontally. We no longer have any means of defining what we mean by space-time. In some sense, therefore, the universe could become smoothed out below the Planck scale.

This smoothing is shown schematically in Figure 12.5. The result is that there are no sharp points in the diagram. The apex of the cone in Figure 12.3, which represents the big bang singularity of infinite density, is effectively washed away.

The precise shape of the universe below the Planck scale is not restricted. We are free, within the context of the theory, to choose any shape that we believe might be appropriate. In 1983, a proposal for the shape of the quantum universe below the Planck scale was put forward by Jim Hartle, from the University of California, Santa Barbara, and Stephen Hawking. Hartle and Hawking propose that the universe be as simple as possible. They suggest that the space-time resemble a *sphere* below the Planck length. In this way, they ensure that the universe has no origin, in the sense that it has no edge or boundary. The universe

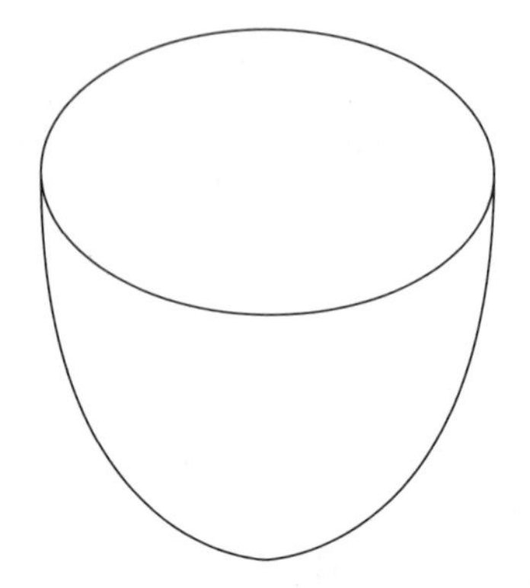

Figure 12.5. The quantum universe below the Planck scale. Space and time lose their familiar physical meaning, and the tip of the cone is smoothed away. The singularity is removed, and the universe resembles a bowl rather than a cone.

may have a finite age, however, since time as we know it today has not always existed.

It is helpful to consider a two-dimensional analogy. The surface of a kitchen table is finite, but it has edges. If we were to roll a marble along the table, it would soon fall off onto the floor. We can define these edges as the region in space where the table begins. Another example of a finite, two-dimensional space is the surface of a sphere. This space is finite in the sense that it would take an ant a finite amount of time to crawl around the equator, but the ant would not fall off the surface as it went on its journey because a sphere has no edge. Thus, there is no point on the surface that we can identify as the origin of the sphere. All points on the sphere's surface are equivalent.

We may summarize this quality by saying that the surface of a sphere has no boundary. According to Hartle and Hawking, the same should be true for the universe as a whole. That is, the universe is finite but has no boundary. For this reason, we cannot associate a particular point in Figure 12.5 with the origin of the universe in the same way that the tip of the cone in Figure 12.3 represented the origin before the quantum fluctuations were considered.

The quantity that we measure today as time does have a beginning in some sense. When the universe was smaller than the Planck limit, there was no such thing as time. When the size of the universe was roughly 10^{-35} metres, the intrinsic quantum fluctuations associated with space and time became negligible. At that point, space and time began to take on separate identities, and the concept of time became meaningful. The origin of time in this picture may therefore be identified as the point where this transformation occurred. In this sense, the universe is not infinitely old, and time has not existed forever even though there is no boundary or edge to the time dimension. If we were to follow the time dimension backwards, we would find that it combines with the other space dimensions to form a smooth, *closed* surface.

Although the no-boundary proposal of Hartle and Hawking is very attractive, it should be emphasized that there is currently no experimental evidence to indicate that it is correct. Recently, another picture for the origin of our universe has been developed by researchers, in particular by Lee Smolin of the Pennsylvania State University. This picture is based on the physical processes that might occur inside a black hole. As seen in Chapter 11, a black hole forms whenever matter collapses so much that light is unable to escape from its gravitational pull. For example, a very massive star may collapse into a black hole when the nuclear fuel inside the star's core becomes exhausted. Alternatively, tiny black holes may have formed throughout the universe immediately after inflation.

According to Einstein's theory, the force of gravity is so strong inside the event horizon of the black hole that nothing is able to prevent further collapse. The theory predicts that the matter should collapse unhindered until its density becomes infinite. This means that the volume occupied by the matter should ultimately vanish. A singularity is said to have formed when this point is reached.

We have already seen that Einstein's theory does not account for quantum fluctuations in the force of gravity. For this reason, it is unreliable over scales smaller than the Planck length. We cannot employ it to investigate what happens to the collapsing matter once its density has become too high. Quantum gravitational effects need to be incorporated into the picture once more if we are to understand what eventually happens to the matter when it collapses into a black hole.

We have discussed how the very notions of space and time lose their physical meaning on scales shorter than the Planck scale. We simply cannot talk about space and time on scales smaller than this. The same is true for the curvature of space-time. Arguably, if we wish to discuss these concepts in the traditional manner, we must limit ourselves to scales that are comparable to the Planck scale. It is then reasonable to suppose that this would limit the curvature of space-time in such a way that it could not exceed a certain critical value. Moreover, since the curvature is intimately related to the distribution of matter, the matter density in a given region should be restricted in a similar way.

The question we are addressing here is whether quantum fluctuations are able to halt the collapse by preventing the density from exceeding an upper limit. There are hints in superstring theory to suggest that the collapse might be halted, and it is possible that a superstring cannot be localized within a region smaller than the Planck length. If this were the case, the formation of a singularity would no longer be inevitable. A definite answer to this question is far from our reach at present. Nevertheless, we may treat this possibility as an hypothesis, albeit a somewhat speculative one. We may then consider also the consequences of adopting such an approach. If it leads to interesting physical results, we would have strong motivation for considering it in further detail.

Suppose, therefore, that the collapse is indeed halted when the density reaches that associated with the Planck scale. What would happen next? It is likely that the effects of the superstring theory would be important at this stage. We have discussed in earlier chapters how this theory can lead to an epoch of inflationary expansion. One possibility, then, is that the conditions inside the black hole might lead to inflation. Since inflation causes space to expand very rapidly, a new era of expansion may follow the contraction.

We have immediately encountered a paradox. If the space inside a black hole inflates, its volume will increase by a huge factor in a very brief interval of time. The final volume of space could be larger than our observable universe if enough inflation occurs. On the other hand, nothing inside a black hole can ever get out, so any expansion that does take place must be confined to the interior of the black hole. The problem arises when we attempt to find room for all this inflated space. The event horizon of a typical black hole is relatively small and is certainly not as big as the entire universe. How is it, then, that a region of space as large as our observable universe can be contained within such a black hole?

We may understand how this paradox is resolved by considering a two-dimensional analogy. Let us return to the ant that can move around on a two-dimensional surface such as that of a balloon. In this analogy, the balloon's surface may be thought of as the space in the universe.

The ant will interpret a black hole in its universe as a closed region of the surface that it cannot see. By walking carefully around the event horizon of this black hole, the ant will conclude that it consists of a finite area of elastic. Once the matter inside this black hole has collapsed

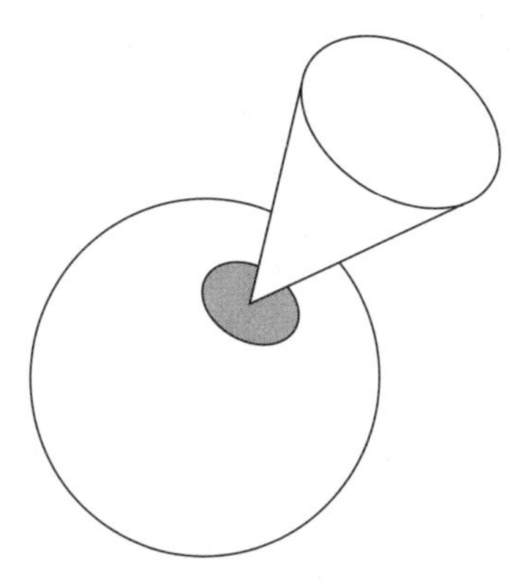

Figure 12.6. An ant located on a balloon is only aware of the two dimensions associated with the surface. Such a being would interpret a black hole as a shaded region on this surface. Events occurring within such a region are shielded from the ant. A new universe, as represented by the cone, may be generated inside the black hole by stretching the elastic of the balloon inside the event horizon in an appropriate way. The size of the cone is not restricted by the size of the black hole because it expands in a direction perpendicular to that of the balloon's surface.

sufficiently, inflation may ensue, as we discussed earlier. This will cause the elastic of the balloon inside the event horizon to stretch. The elastic does not expand back into the two dimensions of the ant's world. Instead, it moves upwards in a direction that is *perpendicular* to the surface of the balloon. This is shown in Figure 12.6.

This explains why the extra space created by the inflationary expansion appears to be contained within the black hole. There is no limit to the extent of the third dimension, so there is more than enough room for the newly stretched elastic. The ant is not aware of this extra space, because it cannot measure it. The ant can account only for the two dimensions associated with the surface of the balloon.

A similar process could occur in the real universe. This implies that the expanded space inside the black hole will behave as an inflating universe in its own right. Since it has been produced from the formation of a black hole, it may be viewed as a 'baby' universe. We may then think of the universe in which the black hole originally forms as the 'mother', because it produces the baby. The relationship between the mother and baby is shown in Figure 12.7. The two universes are connected by a tube of space-time, which, in some sense, plays the role of the umbilical cord.

We saw in the preceding chapter that quantum fluctuations in the region just outside a black hole cause it to lose mass and effectively emit

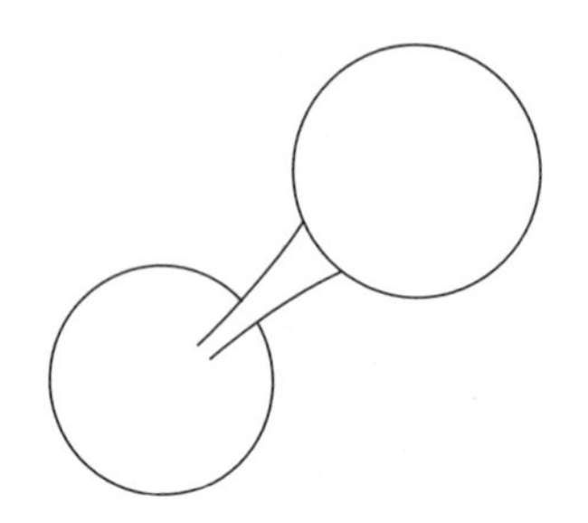

Figure 12.7. A baby universe on the right is produced from the mother universe. The two are connected by a tube of space-time.

particles. The event horizon of the black hole steadily shrinks. A fundamental question currently under investigation is what happens to the black hole once its event horizon has shrunk down to the Planck size. Some researchers maintain that the black hole will stop evaporating at this point because of quantum effects. An alternative view is that the evaporation will continue until the black hole disappears altogether. In this case, the event horizon will effectively shrink to zero.

If this second option is correct, it would have important consequences for the scenario depicted in Figure 12.7. The umbilical cord connecting the two universes probably has a diameter comparable to that of the Planck length. Its point of contact with the mother universe is located within the event horizon of the black hole. If this horizon were to shrink below the Planck length and eventually vanish, there would be no room left for the umbilical cord.

Where would the baby universe go? One possibility is that it could break away completely from the mother universe. The mother and baby would become disconnected from each other, and the baby universe would begin to behave as a separate entity. In this picture, therefore, the formation and subsequent evaporation of a black hole would lead to the birth of a new universe.

What about the development of the baby universe once it has been born? If it is initially inflating, the physical processes in operation will be similar to those that we discussed in Chapter 9. Eventually, some regions of the baby universe will stop inflating, and a standard, hot big bang phase will take over.

The quantum fluctuations that are inherently present during inflation will lead to random irregularities in the distribution of matter. As we discussed towards the end of Chapter 11, these irregularities will be so large in some regions that tiny black holes will be formed. These black holes will be many times smaller than those that form out of collapsing stars. Even so, some will still be fairly massive and will be able to survive for billions of years. Others, however, may be as light as a few grammes. These black holes have very short lifetimes and evaporate within a few seconds, but the space inside them will already have inflated by the time the evaporation is completed.

The process that we have described above will therefore be repeated. New universes will be produced inside the evaporating black holes that are formed when the original baby universe stops inflating. In a sense, these universes will represent the 'grandchildren' of the original mother universe. Since the universes belonging to this second generation will also undergo inflation, more black holes will be produced when their inflationary expansion ends.

We are led to a new picture of the universe. This is shown in Figure 12.8. We shall refer to this picture as the *global universe*. The global universe consists of a network of closed baby universes. Some of these are connected to each other via black holes that have not fully evaporated. Others will be isolated from their parents because the original black hole in which they formed has completely evaporated. New black holes are produced in most of these babies, which results in the next generation of universes. These in turn inflate and produce new black holes.

The key point is that the original baby universe is able to *reproduce*. It grows to become a mother in its own right, and the same is true for its children and their children and so on. The cycle of generating

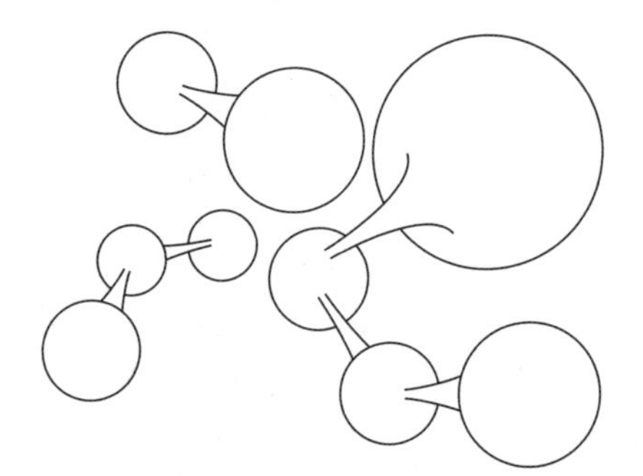

Figure 12.8. The global universe consists of a network of baby universes that are generated inside black holes. These baby universes inflate, thereby producing more black holes and more baby universes. The quantum evaporation of the black holes allows the baby universes to separate from one another, and our universe may have been created in this way. In principle, this process can be repeated indefinitely.

new, inflating universes inside small, evaporating black holes is self-perpetuating. Once it has started, it can continue indefinitely. The global universe may never die and may produce baby universes into the infinite future.

The final size of a baby universe is determined by the amount of inflationary expansion that it undergoes when it first forms. This in turn is determined by the conditions that arose inside the black hole. These will vary from black hole to black hole, so the baby universes will inflate by different amounts.

Some will inflate only for a very short time. This means that they will soon begin to recollapse and will never survive long enough for life to develop. Others may never stop inflating, and their expansion will be too rapid. Since baby universes are continually being produced in this picture, eventually a point will come when the inflationary expansion inside one of them lasts for just the right amount of time for stars to form. We inhabit such a baby universe. It is one that contains a substantial number of stars, one of which happens to be our sun.

The inflationary expansion of our baby universe enabled it to grow to a very large size, but how did our universe become so massive? It is at least 10^{22} times more massive than the sun. Where did all this mass come from?

We might expect the mass of the baby universe to be restricted by the amount of matter that falls into the black hole, but this is not the case. We have seen how mass can be created 'from nothing' as long as an equal quantity of negative gravitational energy is also produced. In this way, the total amount of energy is conserved. There was no energy beforehand and there is none afterwards; the negative gravitational energy precisely cancels the positive energy associated with the mass. Hence, the mass of the baby universe can be quite large. Indeed, the baby can easily grow to be more massive than its parent.

In this scenario, our universe was created when the space inside a newly formed black hole began to inflate. This implies that we could be occupying the inside of a black hole at the present time. When we were discussing the properties of black holes in Chapter 11, we emphasized that one can never see directly inside them. It seemed that the inside of a black hole represented forbidden territory. The opposite may be true in the picture we are now discussing. In this case, the whole of our observable universe is contained within a black hole. This opens up the possibility that we may be able to investigate what lies inside a black hole simply by studying the structure of our own universe.

Our baby universe had an origin in the sense that it came into existence when the black hole that spawned it first formed. It is reasonable to suppose that our universe did not exist before the black hole. On the other hand, it is not clear whether the global universe itself had a definite beginning. If the process of self-reproduction does not stop once it has started, it is possible that it has always been occurring. One could argue that the global universe has always existed in this self-reproducing state. In principle, the global universe may not have had an origin.

This is just one possible interpretation. The global universe might still have been created from some vacuum fluctuation via the process we discussed at the beginning of this chapter. Alternatively, it may be completely self-contained in the sense envisaged by Hartle and Hawking.

To conclude, let us summarize what we have discussed in this book. In this chapter, we have considered two scenarios for the origin of our universe. The first was based on a proposal developed by Hartle and Hawking, where the universe has no definite boundary. The second revolves around the idea that the formation of a black hole may result in the generation of a new universe. This baby universe may have a finite age, but the global universe need not have a well-defined origin.

Our current understanding of the history of the universe is that the superstring theory applied when the universe was just 10^{-43} seconds old. In the chaotic inflationary picture, different Planck-sized regions had different initial conditions. In some regions the conditions were suitable for inflation, but the universe need not necessarily have been hot before inflation began. Those regions that inflated underwent a very rapid expansion and increased in volume by a huge factor. This inflation of the universe can explain, at least in principle, why the universe is so large today and yet contains stars and galaxies.

When inflation came to an end, there was a huge transfer of energy. The energy that had been driving the inflationary expansion was converted into elementary particles and radiation, which resulted in a dramatic increase in the temperature of the universe. It is probable that the temperature would have exceeded that required for the electroweak force to operate. Eventually, the temperature fell sufficiently to cause the electroweak force to split into two separate components. These are identified today as the weak and electromagnetic forces. This separation was completed about 10^{-10} seconds after the end of inflation.

By this time, the temperature had dropped to 10^{15} degrees. The quarks were so tightly pressed together that they could not feel the confining influence of the gluons. They effectively behaved as free particles. Their average separation increased along with the cosmic expansion, and, after about 10^{-4} seconds, the quarks became trapped into pairs or triplets. These bound states rapidly decayed, and the only particles comprised of quarks after this time were the neutrons and protons.

The neutrons remained free until about three minutes had elapsed. By this time, the temperature of the universe had dropped sufficiently for the neutrons and protons to bind together to form nuclei. During this process of 'nucleosynthesis', conditions were changing very rapidly. There was only sufficient energy and time available to form the lightest nuclei. Many of the protons remained free and eventually went on to form hydrogen. The neutrons and the remainder of the protons combined to form helium and a small quantity of other elements.

The expansion continued after nucleosynthesis, but nothing significant happened for a further three hundred thousand years. The energy of the photons remained high enough to prevent the electrons and nuclei from forming atoms. After this time had elapsed the photons had lost much of their energy in the expansion. The electrons and nuclei were then free to combine into neutral atoms. Since radiation is not affected by electrically neutral matter, the former became essentially free from the latter. The universe became transparent in this era.

Gravity was now the dominant force in the universe. The tiny primordial fluctuations generated during inflation grew in size under the influence of gravity. The universe became increasingly lumpy, and islands of relatively dense matter gradually developed throughout the universe. These islands were not precisely uniform, and they fragmented into many separate mini-islands.

The temperature of the matter in these mini-islands increased as they collapsed. Their centres become so hot that hydrogen nuclei were able to fuse together to form helium. This conversion of hydrogen into helium released enough energy to prevent further collapse of the mini-islands, and they developed into stars.

Typically, a star that converts hydrogen into helium can survive for billions of years. Generally speaking, more massive stars burn hydrogen more efficiently. Many of the stars that formed soon after the big bang may have been quite massive. These had shorter lifetimes and underwent further collapse when they ran out of hydrogen. The helium nuclei in the core then fused together and formed the carbon, nitrogen and oxygen that is so essential for life here on Earth.

The massive stars finally exploded as supernovae when their outer regions were blown away. Our solar system is thought to have been formed from some of the stellar material that was released in one of these explosions about five billion years ago.

It should be emphasized that although these new ideas regarding the state of the universe before the Grand Unified era are certainly appealing from a physical point of view, we currently have no direct observational evidence that either verifies them or rules them out. The temperature irregularities in the cosmic microwave background radiation that we discussed in Chapter 9 do provide strong support for the idea that these fluctuations were generated at, or shortly after, the GUT transition. The observations are not sufficiently accurate at present to allow us to conclude for certain whether the fluctuations were generated by an inflationary expansion of the universe. However, we will be able to test the idea of inflation within the next few years as the quality of the observations improves.

At the start of this book, we went on a rapid tour of what is contained within our observable universe. We encountered problems in dealing with the large distances involved even before we had left the confines of the solar system. We concluded that the observable universe is at least ten billion light years in diameter, and this is indeed very large when compared to our everyday experiences. We now see that this is actually a very small distance when compared to the typical scales that are possible in the inflationary universe. The conclusion we should draw from our cosmic journey, therefore, is that the 'bang' may have been much *bigger* than we had previously imagined.

Index